YOUNG ADULT SERVICES
STO

ACPL ITEM
DISCARDED

Y0-BSM-571

Y 660.6 OL2M
OLEKSY, WALTER G., 1930-
MIRACLES OF GENETICS

2296438

**DO NOT REMOVE
CARDS FROM POCKET**

ALLEN COUNTY PUBLIC LIBRARY

FORT WAYNE, INDIANA 46802

You may return this book to any agency, branch,
or bookmobile of the Allen County Public Library.

DEMCO

MIRACLES OF GENETICS

SCIENCE & TECHNOLOGY

BY WALTER OLEKSY

CHICAGO

Editor: Beatrice Beckman
Designer: Dolores Hollister

Allen County Public Library
Ft. Wayne, Indiana

Picture Acknowledgments
AP/Wide World—10, 13, 15, 16, 17, 20 (bottom), 27, 50, 53, 66, 69, 88, 89, 96, 104, 113, 115
The Bettmann Archive—4, 73
Cincinnati Zoo / © Ron Austing—43
Colorado State University / © Ken Williams—47 (6 photos), 49, 61, 99
Design Plus—6
Howard Frank Collection—102 (2 photos)
Tony Freeman Photographs—64
Genex Corporation—12 (2 photos), 22 (top), 68 (2 photos)
Historical Pictures Service, Chicago—67
Hoffman-La Roche Inc.—11, 77, 105 (left), 110, 122, 124, 125
IBM Corporation—21
Journalism Services: © H. Rick Bamman—26 (top); © John Patsch—82 (bottom); SIU—8 (bottom), 74 (left), 91, 93
Louisville Zoo / 44; © Nancy Scheldorf—45
Medichrome / The Stock Shop: © Arthur Sirdofsky—14, 70 (right), 75; © Joel Landau—20 (top); ©Leonard Kamsler—22 (bottom), 80; © Ashvin Gatha—cover (top right), 55; © Michael Philip Manheim—57 (left); © Mike Yamashita—57 (right); © D.C. Lowe—74 (right); © Alexander Tsiaras—82 (top), 84, 85, 86 (2 photos); © D.J. Heaton—94; © Tom Tracy—11 /; 72

Monsanto Company—118 (2 photos), 119 (2 photos), 121 (3 photos)
The Morton Arboretum—36
© Marc Nadel—28
National Seed Storage Laboratory—cover (top right), 37, 38 (2 photos), 39 (6 photos)
Nawrocki Stock Photo: © Jeff Apoian—24
© News Group Chicago, Inc. / Nancy Stuenkel / Chicago Sun Times—35
Photo Researchers, Inc.: © Bill Longcore—8 (top), 9; © Tom McHugh—19 (top), 63; © L & D Klein—26 (bottom); © R. Van Nostrand—40 (2 photos); © Omikron—70 (left); © Hank Morgan—71 (left), 105 (right); NIH / Science Source—71 (right); NCI / Science Source—101; © S. Stammers—111; © Dr. C. Chumbley—114
Dr. Steven Ruzin/Plant Genetics, Inc.—108
Schmidt Nursery, Boring, Oregon / © Keith Warren—cover (left), 23
Courtesy Dr. David A. Shafer—59
Tass from Sovfoto—18
Tom Stack & Associates—7, 19 (bottom)
R.L. Brinster & R.E. Hammer, School of Veterinary Medicine, University of Pennsylvania—42
Agricultural Research Service, USDA—30, 33, 56

Library of Congress Cataloging-in-Publication Data

Oleksy, Walter G., 1930-
  Miracles of genetics.

  (Science & technology)
  Includes index.
  Summary: Introduces genetic engineering and describes its practical applications in the creation of superior plants and animals and improved human medicine.
  1. Genetics—Juvenile literature.
  [1. Genetic engineering. 2. Genetics] I. Title.
  II. Series: Science & technology (Chicago, Ill.)
  QH437.5.044 1986   660'.6   85-30852
  ISBN 0-516-00531-6

Copyright © 1986 by Regensteiner Publishing Enterprises, Inc.
All rights reserved. Published simultaneously in Canada.
Printed in the United States of America.
1 2 3 4 5 6 7 8 9 10 R 95 94 93 92 91 90 89 88 87 86

TABLE OF CONTENTS

1. Designer Genes... 4
2. Superplants... 24
3. Superanimals... 40
4. Genetics in Humans... 64
5. Test-Tube Babies... 82
6. Can a Human Be Cloned?... 94
7. Genetic Engineering and Cloning—Good or Evil?... 102
8. Today and Tomorrow in Genetics... 108
   Glossary... 127
   More About Genetics... 127
   Index... 128

WOMAN GROWS NEW LEG!

TEST-TUBE BABY BORN!

UNLIMITED CHEAP ENERGY!

DESIGNER GENES
1

History tells us that Dr. Faustus (also known as Faust) was a German magician who lived in the sixteenth century. The historical Dr. Faustus died about the year 1540. Legend tells us that Dr. Faustus had sold his soul to the devil in exchange for youth, knowledge, and power. It is the legend that lives on.

One of Dr. Faustus's dreams, it is said, was to create a human being in a test tube. That dream—whether legend or fact—has become a reality. Modern research and experiments in genetics have turned many such dreams into realities and others into possibilities. Imagine what the legendary Dr. Faustus would think if he could see the work being done in genetics today!

WHAT IS GENETICS?

Genetics is the scientific study of life—life in plants, in animals, and in humans. The dictionary tells us that the word *genetics* means "pertaining to or influenced by genesis or origins." The first book of the Bible—Genesis—is about the creation of the world and man.

Biologically, genetics is the scientific study of heredity. Heredity refers to the way every living thing—from a rose to a puppy dog to a human being—inherits from its parents or species the characteristics that make it uniquely of its own kind. Because of the miracle of heredity, every boy has all the genes to make him look like and be a specific boy, and every girl has all the genes to make her look like and be a specific girl.

Left: The Faust dream of creating *homunculus* (dwarf man) and some realities and dreams of modern genetics

INTRODUCING GENES

A gene is the part of a chromosome that contains the genetic information for a particular trait or characteristic. Genes are the chemical blueprints for all living things. They determine whether an organism will grow into a plant, an animal, a bacterium, or a human being. They also determine sex and the color of the eyes and hair. It is genes that determine who will have a beautiful singing voice, who will be a "natural" athlete, who will be a mathematical genius, and who will have other skills.

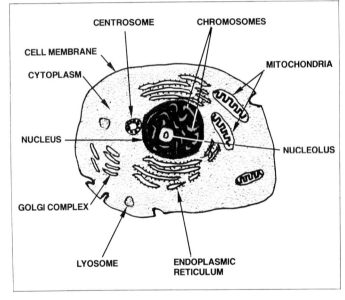

The cell is the basic structural unit of the body. This diagram shows a typical cell, including some of the sites of important events. The nucleus, or center, is where DNA is stored.

Genes do their work inside the cells, which are the tiny building blocks out of which our bodies are constructed. If you look at a cell under a regular microscope, you see only a shadowy blob, a colorless bunch of chemicals. Magnified, you can see that the cell has an information center comprised of the nucleus, the protein-manufacturing units, and the energy-production points.

The genes are in the cell's nucleus (center). They contain all the information the cell needs to carry on its protein and energy production. Looking at the genes much closer and under greater magnification, you will see long, spiraling double strands of atoms. These are called DNA (deoxyribonucleic acid), the master chemical of genes. The sequence or layout of these atoms contains all the information the cells need in order to function.

This computer-drawn simulation shows twenty base pairs of DNA.

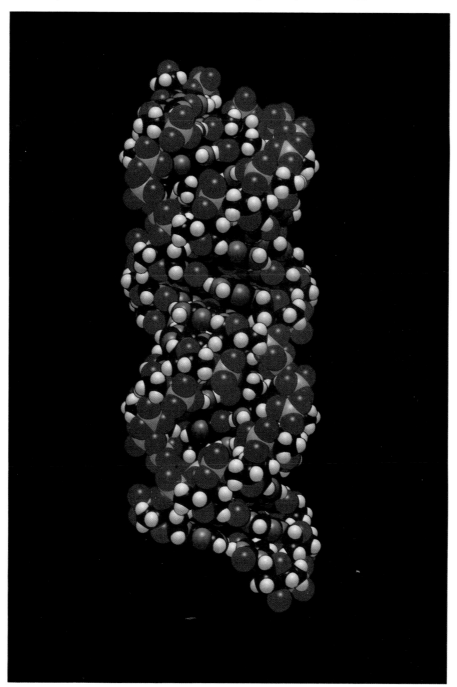

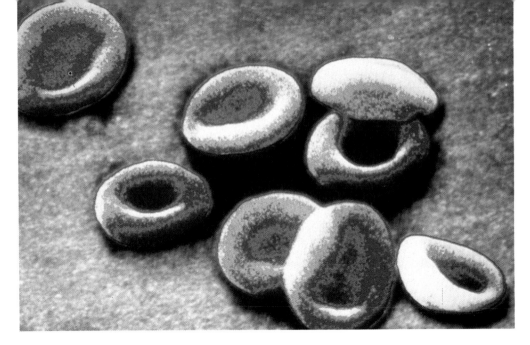

This is a greatly enlarged picture of normal human blood cells. Note the rounded shape.

CHROMOSOMES

Chromosomes are structures in the nucleus of a cell and are composed of protein and DNA (the life force). Man's forty-six chromosomes contain about a hundred thousand genes, each a different sequence of four chemical units. They contain all the genetic information of a cell.

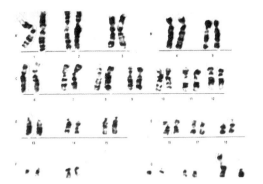

This magnified picture shows human chromosomes. The chromosomes are in the nucleus of the cell and contain all the genetic information of the cell.

Of the approximately 100,000 genes found tucked inside a human cell, about 21,500 now have been traced to their chromosomal locations. Scientists have isolated these genes and identified the specific jobs they do in the body. For example, scientists know the chemical structure of the genes that make insulin, a hormone whose lack causes diabetes, a potentially fatal disease.

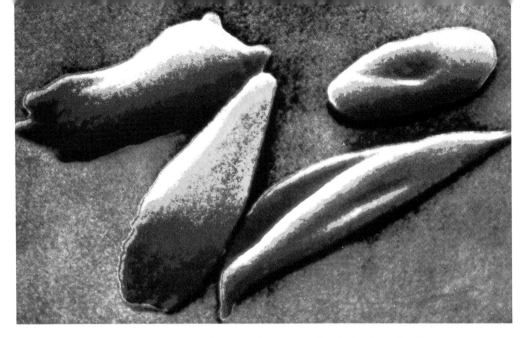

This photograph shows how sickle-cell anemia affects the blood cells. Compare these with the normal cells on the facing page.

CHANGING GENETIC STRUCTURES

New genes are being identified at a rate of about two hundred per year. Scientists are hopeful that by the end of this century, the entire genetic structure of humans will be known.

Although scientists have known about genetics and heredity for hundreds of years, it is only in recent years that they have begun to understand how genetics and heredity work. Today, scientists are studying ways to make changes in the genetic structure of plants, animals, and even human beings. This is called genetic engineering. The idea of genetic engineering is to improve existing life forms—to make them healthier, live longer, be more productive—or to create new, superior strains of a life form.

RECOMBINANT DNA

Genetic engineering—also called gene-splicing—is achieved by recombinant DNA. This laboratory-produced DNA results from a process whereby two DNA molecules from different sources are combined to alter a life form or create a new one. Usually, a piece of DNA from one species is attached to bacteria of another. The word *recombinant* can be thought of as a scientific way of saying that the life-force characteristics are recombined to alter a life form.

It might be said that scientists today are using genetic engineering to create "designer genes." What was considered science fiction little more than a decade ago is now happening in genetics. Using gene-splicing and other genetic-engineering techniques, scientists have developed the means of creating new life or changing evolution.

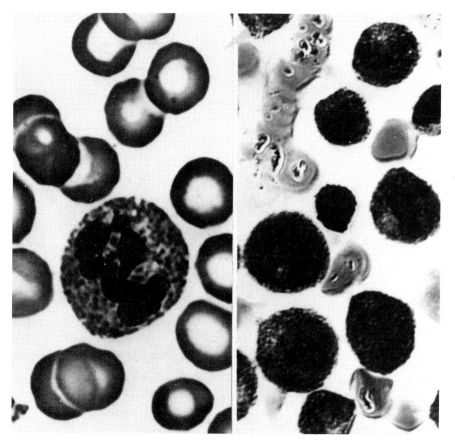

Normal blood cells (left) and leukemia-diseased blood. The abnormal cells of leukemia, unless controlled, multiply and carry cancer throughout the blood system.

By manipulating genes, bioscientists have begun tapping a seemingly limitless new resource. They can remove certain genes from one species and insert them into the genetic machinery of another species to make organisms that never existed before. One scientist says, "Genetic engineering techniques are so powerful, we

Scientists at work in a recombinant DNA laboratory. Gene-splicing and other genetic engineering techniques promise great advances in medicine, agriculture, and other areas of life.

can make micro-organisms to solve a lot of the world's problems, from energy and pollution to hunger."

WHAT GENETIC ENGINEERING CAN DO

The production of recombinant DNA, though only about fifteen years old, has already become almost routine. Bioscientists predict that one day, bacteria will be turned into living factories that will produce huge amounts of vital medical substances such as serums and vaccines to fight diseases ranging from hepatitis to cancer—and perhaps will even help one day to cure the common cold.

Among the studies under way now is one to determine the hormone that triggers growth in humans. Research in this area will allow thousands of children with growth disorders to develop normally. It may also help in the treatment of burns, bone fractures, and diseases of older people.

Recombinant DNA research will help scientists to identify all the genes in the human cell and use their knowledge to replace defective genes with healthy ones. This could enable researchers to find cures for such genetic diseases as hemophilia, a disease that can cause victims to bleed to death, and sickle-cell anemia, which dangerously weakens the body, particularly the legs. With genetically engineered human insulin already on the market, scientists are now testing vaccines against hepatitis, malaria, rabies, and venereal diseases. A cure for cancer also may come out of genetic research.

Also being studied are special enzymes that could turn solid waste into useful sugar and alcohol. Other microbe tests may separate valuable metals from ore and extract and purify oil from depleted wells. As much as half of all industrial chemical stocks may one day be made biologically.

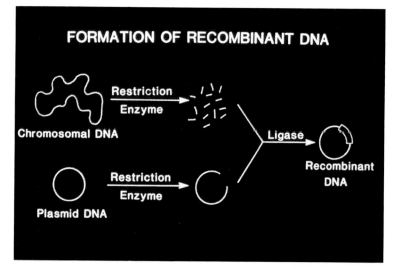

A schematic showing the formation of recombinant DNA

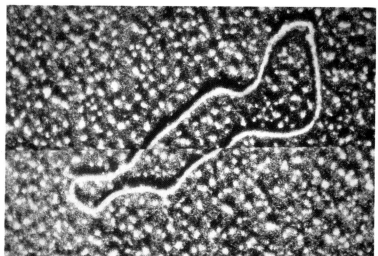

DNA as seen through a microscope

IMPROVING FOOD SUPPLIES

Much of the work of genetic engineers today is focused on improving strains of plant and animal life to create healthier and more productive stock that could help feed the world's hungry. Geneticists already have found that it would be possible to produce a supercow, a dairy cow as big as an elephant and capable of giving forty-five thousand pounds of milk a year, compared with about fifteen thousand pounds for a typical animal.

By growing plant cells in laboratory dishes and transplanting genes between animals or plants, bioscientists are hopeful they can one day produce crops that need no fertilizer, enable pigs and cattle to grow twice as fast as they do today, produce animals with leaner meat, create more nutritious grain, and enable plants to resist disease, drought, insects, and herbicides. Supercows, -corn, -rice, -grain, -beans, and other livestock and plants could play major roles in feeding the hungry of the world.

Genetic engineering is not, however, without its critics. Many religious leaders and others are opposed to tampering with the genetic code by modifying genes. They fear that genetic engineering may be as dangerous to the future of humanity as are chemical, bacterial, or nuclear weapons.

Both chickens in this photo are eight weeks old. The size of the smaller one was arrested by withholding an organic chemical known as tryptophan. The chick's life is extended by the same length of time its growth is stopped. When its regular diet is restored, the chick will develop normally.

TEST-TUBE PREGNANCIES

Early in 1980, the first laboratory to create test-tube babies in the United States was approved for the Eastern Virginia Medical School in Norfolk, Virginia. Test-tube babies already had been successfully born in England and India, bringing new hope to thousands of women who were infertile and unable to bear children by traditional methods. Today, there are several dozen test-tube babies in the world, and the practice of making women pregnant by artificial methods is growing.

Sex engineering has already revealed some startling information, such as that everyone is perhaps basically female when conceived. Scientists investigating the age-old puzzle of what determines maleness or femaleness have concluded that nature has an almost overpowering tendency to want to make all babies female. If it were not for a recently discovered molecule, called the "ultimate determinant of maleness," which is sometimes added to the embryo several weeks after conception, all babies would be girls. Scientists call this the Eve principle, and its discovery is part of a major revolution going on today in the fields of embryology and genetics.

Before so-called test-tube babies were possible, this couple would have remained childless. In vitro fertilization has brought hope to numerous childless couples.

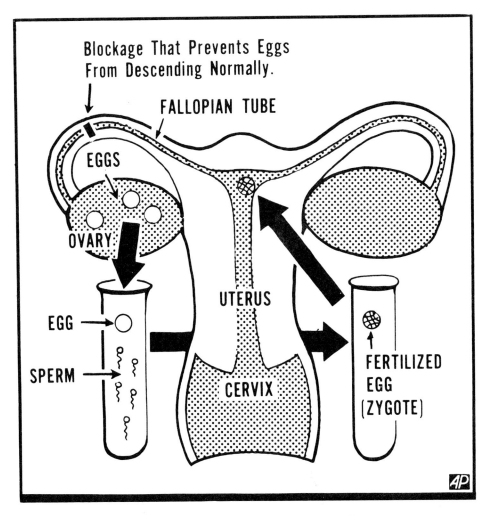

The first test-tube baby was born to Mr. and Mrs. Gilbert John Brown of England. This drawing shows how the baby was conceived outside Mrs. Brown's body. A blocked fallopian tube (top left) had prevented normal conception.

NEW GENETIC MARVELS

Japanese researchers into genetics have introduced a new plant miracle—a square watermelon. It tastes about the same as an oval melon, but its shape makes it easier and more economical to transport in large quantities.

This young woman is holding two square watermelons. The melons were developed by Tomoyuki in Japan.

Hybrid flowers and vegetables have been created in laboratories and greenhouses for years, producing healthier and more colorful forms of flowers and larger, more nutritious, and disease-resistant vegetables. Only recently, we were introduced to the lem'n lime. On the outside, it is green; on the inside, pale orange. It looks like a fat lime and tastes like a juicy, sweet lemon.

A few years ago, lem'n limes didn't exist. Now thousands of pounds of lem'n limes are shipped from Florida's citrus groves to stores all over the country each year, and we enjoy a new fruit and drink. Lem'n limes evolved because some lemon roots got tangled with a Persian lime tree in a chance occurrence. Out of the accident, the lem'n lime was born.

This lemon, grown by Robert Jennings of Downey, California, weighs five pounds and is eight inches tall and twenty-three inches around. Genetic engineering could produce such superfruits consistently.

The American eagle, among the most endangered of birds, may be bred in larger numbers in captivity thanks to the work of geneticists and the St. Louis (Missouri) Audubon Society. They have begun successfully hatching eagles by artificial insemination.

A few years ago, American frogs were in very short supply for laboratory experimentation. Researchers immediately stepped up their attempts to clone frogs. Geneticists already were successful in cloning tadpoles. The frog shortage has stimulated advanced cloning research, even though critics are fearful that such research may lead to cloning humans.

SOME DANGERS OF GENETIC ENGINEERING

Some critics of cloning are fearful that instead of creating duplicates of the healthiest, most intelligent human beings, geneticists, either

If geneticists succeed in their cloning efforts, this long-extinct mammoth may live again. Beside the replica of the adult mammoth is the figure of the corpse of a baby mammoth.

by accident or for some evil purpose, may bring into the world clones of monsters such as Frankenstein or Adolf Hitler.

Geneticists try to assure us that they have no such intention of releasing into the world genetic misfits or evils of any kind. They reassure us that through a variety of often surprising experiments and investigations, they are working for a better world partly by studying the past and adapting new research to long-dead species.

California insect pathologists recently began studying a fungus gnat believed to be forty million years old. Its nuclei and other small cell structures were remarkably still intact. Geneticists hope to acquire DNA from the gnat's cells and perhaps mate it with a modern fly to give birth to the insect's ancient ancestor. As with cloning a mammoth, it is part of geneticists' efforts to resurrect long-vanished life forms, which can lead to revolutionary new discoveries about plants, animals, and humans who have survived the evolutionary process since the world began.

MEDICAL MIRACLES

Because of genetic engineering in plants and animals, scientists believe they may someday be able to enable humans to grow new parts of the body, such as arms or legs. Some lower animals, such as crabs, can grow new claws and some lizards can grow new tails. The process is called regeneration.

Some regeneration already occurs in humans. We naturally grow new skin, hair, bone, portions of the liver, cells lining the digestive tract, and blood cells. But so far a lost limb can only be replaced with an artificial one. However, medical miracles have been worked in implanting into humans artificial organs such as kidneys and hearts. Now, some scientists say that for a human to grow new arms or legs isn't as farfetched as many think. They maintain it is just a matter of time and experimentation.

Fetal brain cells from one strain of rat have been transplanted into the brains of adult rats of another strain. The transplanted cells grew and functioned normally, correcting brain defects. Scientists are

Regeneration occurs in some lower animals. This desert iguana (top) is growing a new tail. The starfish (bottom) is regenerating three of its arms. Some geneticists believe that they may someday find ways to enable humans to grow new limbs.

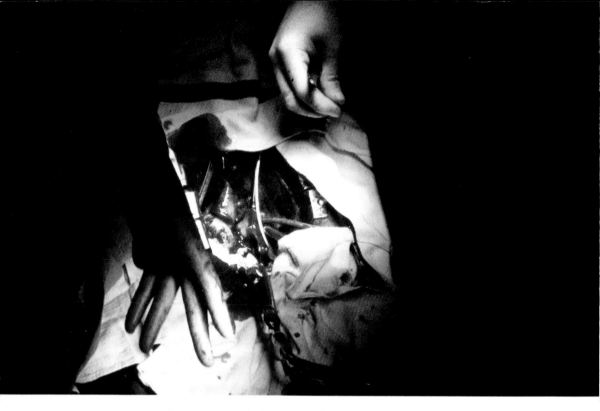

Heart transplants have made it possible for people with failing hearts to live long, normal lives. Other organ transplants also have given new life to seriously ill people.

Dr. Harry Whitaker, a University of Rochester neuropsychologist and neurolinguist, holds a model of a human brain. Dr. Whitaker is doing research into the part of the brain that is involved in language—the left hemisphere.

encouraged that in the not-too-distant future, it may be possible to use cell transplants to rejuvenate aging human brains. The damage of strokes may be reversed, memory and intelligence improved, and other brain defects corrected.

Not too far in the future, a kind of organic computer may be developed if scientists can modify proteins from bacteria to create living "biochips." Computer chips that store information could be constructed from biological materials such as proteins, enzymes, and DNA. Such biological chips—biochips—could perform calculations in about a millionth of the time it takes today's best computer chips, which are made of metal or silicon.

This computer chip is so small that it could fit through the eye of the needle with which it is pictured. Living biochips could one day enable humans to think at computer speed.

The combination of computer science and genetic engineering may one day produce computer implants that could enable the blind to see, could give the human mind a photographic memory, or could enable the mind to work with the speed of a computer.

A HUMAN CLONE?

In 1978, a book was published as nonfiction, claiming that a human boy had been cloned. The book told of a wealthy man who paid a million dollars to have a son cloned of himself and of how genetic scientists actually succeeded in producing the clone. The astounding revelation drew wide attention and skepticism from many scientists and religious leaders. Several years later, the publisher admitted that the story was a hoax and that the book should have been published as fiction, not fact.

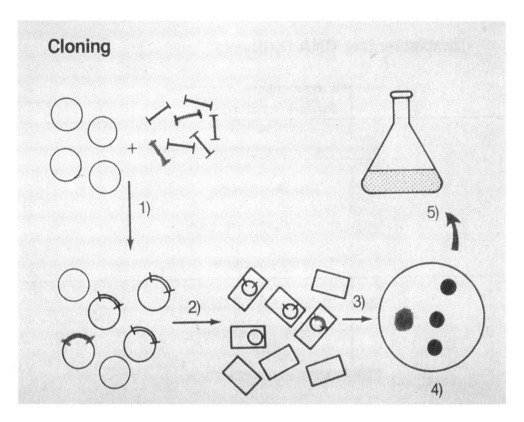

This simplified schematic (top) shows the steps in the cloning process. The photo at the bottom shows cloned tadpoles in petri dishes.

This nursery row of cutting-produced elms represents a single clone. They are, therefore, uniform in their growth.

Yet, scientists do believe it may be possible one day to clone human beings. Superior strains of men and women could be born with great intelligence, bravery, or artistic ability. Critics of human cloning continue to argue that cloning humans might open a Pandora's box, unleashing potentially harmful mutant beings on mankind. The debate goes on as the cloning of a human being becomes a greater possibility.

The world of genetics is as old as the world itself, and yet it is as new as tomorrow because of genetic engineering. Geneticists believe that their work with gene-splicing and other genetic engineering techniques is good and will work for the benefit of mankind, enabling us to lead healthier, longer, happier, and more productive lives.

In the following chapters you will learn more about genetic engineering in plants, animals, and humans. You will discover more about the changes that bioscientists are bringing about in our lives as we look at where research is now and where it is going in one of the most exciting frontiers of science today—the revolutionary world of genetic engineering.

SUPERPLANTS
2

While genetic engineering in animals and humans is still relatively new, plants have been genetically altered and cloned for thousands of years. Since biblical times, the heredity of seedlings has been altered to enable plants to become better adapted to new climates, be more disease-resistant, and grow bigger and more nutritious. Through the technology involving recombinant DNA, many common fruit trees and farm crops have been bred from plants that may today bear only a slight resemblance to their ancestors.

We have noted that the use of recombinant DNA—genetic engineering—involves the transfer of individual genes (units that govern the machinery of living cells) from one organism to another. In simple organisms, scientists usually alter only one gene—or at most a few genes—to achieve useful change. In plants, which include some of the more complex organisms in nature, entire blocks of genes may be altered.

Geneticists working in plant breeding alter the heredity of plants through crossbreeding, which is the transfer of pollen from the male portion of the plant to the female flower. The plant that results from this union is a genetic combination of traits from both parent plants.

EARLY PLANT BREEDING

The American Indians can be credited with one of the first and most amazing plant-breeding successes. Long before the New World was discovered, native American Indians bred ten major types of corn that we know today. Scientists believe that the corn was bred from the wild teosinte plant, which grows in Mexico, Guatemala, and

Left: Genetic engineering may make it possible for lush vegetation such as that found in rain forests to grow in other climates.

The wild teosinte plant of South America is believed to be the ancestor of this modern corn plant.

Honduras. The Indians probably brought the plant from South America and bred it to withstand the climates of the north.

Plant breeders also have been cloning plants for thousands of years. Most fruit trees and wine grapes are grown from clones. When a grapevine is found to be good for producing exceptional wine-making grapes, the branches of the good vine can be grafted onto another rootstock. The resulting grapes are clones—identical copies of the grapes of the good vine.

Most wine grapes are grown from clones. These are cultivated Oregon grapes *(Berberis aquifolium)*.

A University of Rhode Island plant researcher, Dr. William Krul, examines an infant grapevine that was grown from a cloned embryo. In the background are more tiny plants, all of which were cloned by the same parent.

A GREENER REVOLUTION

Although the genetic manipulation of microorganisms in the laboratory has become routine, scientists have just begun to successfully alter the heredity of plants with modern gene-splicing techniques. Bioscientists at the University of Minnesota, who are studying the genes of corn plants, are among researchers who are optimistic that many genetic miracles in the plant kingdom are ahead as recombinant DNA technology is applied to plants.

Plant breeders have already made major improvements in crops since the 1930s, increasing yields in grain crops by an average 2 to 3 percent a year. In the 1960s and 1970s, scientists raised crop yields dramatically through crossbreeding and hybridization. This was called a Green Revolution. Now a new Greener Revolution is predicted, one in which genetically engineered crops will be more nutritious and produce even higher yields.

Genetic engineering could add $5.6 billion to annual crop value before the year 2000, some scientists predict. Biochemists believe they are no more than a few years away from breaking the "gene barrier." They hope to mix the DNA of species that cannot be crossed by conventional breeding to produce valuable new varieties. Perhaps they might thus, for example, be able to give wheat the soybean's ability to extract nitrogen from the air and convert it to fertilizer within its root system.

It will take some doing, however; scientists know far less about the genes of plants than they do about those of bacteria and animals. They also have been held back in their research by lack of money, since most funding in the past has gone to medical research involving areas such as heart transplants and finding cures for cancer. And they have further been hampered by critics who fear that bioscientists working in genetic engineering might endanger the environment. Tampering with plant and animal life, these people say, might alter the delicate balance of nature. A harmful microorganism might accidentally be introduced into the atmosphere.

With the hope of reaping the rewards of genetic improvements in the plant kingdom, some major corporations are starting to invest heavily in bioscientific experiments on plants. Campbell Soup Company is investing $50 million in a program to bioengineer a greater

Companies that produce tomato soup and catsup have hired scientists to produce a supertomato with more solids. Tomatoes are 95 percent water. And that water is costly every step of the way from the field to the packing plant, where most of the fluid has to be cooked away.

tomato for its famous soup. DuPont, General Foods, Monsanto, Standard Oil of Indiana, Union Carbide, and Weyerhaeuser are investing in agrigenetics—genetically improved agriculture.

Monsanto chemical company's new $170 million research station outside St. Louis, Missouri, has growth chambers called microclimes, small experimental replicas of different climates. So far they have simulated conditions from the deserts of Saudi Arabia to the rain forests of Brazil in order to better study how plant life grows in various climates and how food plants can be improved.

PLANT CLONING

Some agrigenetic research projects are more advanced than others. Scientists have become skillful at manipulating cells so they can reproduce identical copies of whole plants through cloning. The U.S. Department of Agriculture (USDA) is using the technique to propagate dwarf apple, pear, and other fruit trees. Such trees will be easier to harvest and will bear fruit sooner than normal fruit trees.

Botanists have already created new products that range from a disease-resistant sugarcane to tomatoes that are less watery and brighter in color. Such tomatoes mean that producers of soups and ketchups don't have to extract so much liquid during processing. The result is a greater yield from fewer tomatoes.

The next step for agrigenetics is still down the road a way. It involves not only exploiting traits that are already in a plant's genes, but also introducing foreign characteristics. For example, in the technique of protoplast fusion, biologists engineer a sort of botanical shotgun marriage. They use enzymes to dissolve the walls around cells from different plants and then mix them in a chemical solution until they fuse. Think of the possibilities of such fusions.

GENE-SPLICING IN PLANTS

Gene-splicing—the use of recombinant DNA—is considered to be the most appealing technology in agrigenetics, even though botanical gene-splicing is the most difficult breeding technique of all. "You're tailor-making your organism rather than mutating what comes from tissue culture," says biologist Winston Brill, a member of Agracetus, a biotechnology firm.

It may be hard to believe, but the basic molecular biology of plants is not as well understood as is the basic molecular biology of mammals or bacteria. String beans, for example, are said to have ten times as much DNA as humans.

Despite the complexity of gene-splicing in plants and the drawback of limited research funding, the results achieved by agrigeneticists so far have been very impressive. Researchers in Monsanto laboratories have used a bit of DNA from one plant virus to carry a gene for antibiotic resistance into petunia cells. The cells grew into whole plants, retained the functioning gene, bore seed, and passed the trait on to the next generation. This minor miracle proved that plants will retain and express an alien gene.

Researchers at the Cell Culture and Nitrogen Fixation Lab in Beltsville, Maryland, work to increase soybean yields by the introduction of highly efficient *Rhizobium* bacteria strains.

Another Monsanto project involves giving plants the gene necessary to make a "living insecticide." This could produce a crop less dependent on chemical poisons, something that ecologists and environmentalists would love.

Biogenetic research is moving forward.

In 1984, Calgene, a biotechnology company in Davis, California, announced that it had taken a step toward the transfer of an agriculturally important trait. The company's scientists had spliced into tobacco, tomato, and soybean cells the gene for resistance to the popular herbicide Roundup.

In still another program, scientists at the University of California at Davis are trying to transfer the dwarfism gene from peach to almond trees. They hope to create trees with more fruit and less lumber.

ISOPENTENOIDS

Meanwhile, the U.S. Department of Agriculture (USDA) points to growing interest in another area of biogenetic research—isopentenoids. These are chemical substances common to all plants and animals, including human beings.

Isopentenoids bind together all living cells. Some of these natural chemicals regulate growth, production of toxins, and reproduction in plants and animals. Some provide organisms with such biologically vital compounds as hormones, pheromones, and many types of lipids, including cholesterol. And some impart flavors, aromas, and nutrients to foods, as carotene does for carrots and tomatoes.

Scientists of the USDA's Agricultural Research Service predict that isopentenoids may someday be harnessed for such varied biological purposes as enabling stunted plant seedlings to survive the stress of drought, protecting plants from predator insects, bolstering nutrients in foods, preventing animal birth defects caused by plant toxins, and hastening crop maturity.

"All living things require isopentenoids," says Dr. Terry B. Kinney, Jr., the USDA research agency's administrator. "As our research helps piece together a fuller understanding of their versatile roles in living matter, we will also learn more about their roles in ecology and evolution."

"The benefits of basic research on isopentenoids go far beyond agriculture," says Malcolm Thompson of the ARS Insect Physiology Laboratory, Beltsville, Maryland. "Information we generate enables chemists, biochemists, nutritionists and others to solve problems in their fields.

"Understanding the basic mechanisms of how plants interact, often concurrently, with insects, mammals, fungi, environment, and other plants is an awesome task. This research challenge is far more complex than focusing only on chemical reactions that occur in mammals. Yet mammals, especially humans, are affected by the increase or decrease in food supply brought about by certain isopentenoids."

So far, most isopentenoid research has focused on five basic interactions: plant/insect, plant/mammal, plant/plant, plant/fungus, and plant/environment.

TISSUE CULTURE

At the heart of most agrigenetic engineering is the relatively new technique of tissue culture. Tissue culture refers to the growing of tiny collections of plant cells in a solution of salts, sugar, vitamins, and hormones. It is possible to regenerate whole plants from these cultures, as commercial orchid growers are now doing. The advantage of using this technique is that the grower has increased control over each plant's genetic makeup.

The combination of tissue culture and gene-splicing has a potential so powerful for agriculture and plant science that it has created an industry perhaps even larger than that designed to transfer human genes to other life forms. One recent study projects a market for plant bioengineering that could reach $100 billion worldwide in the next ten or twenty years.

Efforts to improve the world's food supply are easy to understand when we realize that in the next two decades, food will not only be needed to feed the current 231 million people in the United States, but also about 20 million more. And Latin America might have ten times that increase. By the year 2000, the world population is

Geneticists are hoping to breed sorghum and other plants with traits such as resistance to disease, high yield, and improved seed quality.

expected to have increased by almost half, to just under six billion. But if current rates of land degradation persist, by that time one-third of the world's arable land will have been destroyed.

PLANT MIRACLES

Biologists predict that gene-splicing, tissue culture, and other techniques will enable them to accomplish great things. They anticipate being able to transfer to such food crops as corn, wheat, and soybeans the genes that give some plants natural resistance to herbicides. They feel they will one day be able to transfer or engineer into plant genes resistance to diseases caused by fungi and viruses. A plant strain with such valuable resistance could then be selected and multiplied by using tissue culture techniques.

Researchers indicate that plants could be bred to withstand stressful temperature and humidity levels. They could be made to grow in salty environments. As one researcher put it, "If we could only teach corn what salt hay grass knows!"

As a result of bioengineering techniques, plants could be made to increase their yield, produce more nutritious proteins, self-fertilize, and be more resistant to insect damage. Also new hybrids could be introduced into the plant kingdom, such as the potato-tomato.

Agrigenetics, still in its infancy, has already produced some marvels in the plant kingdom. In 1982, scientists from the Department of Agriculture and the University of Wisconsin created a "sunbean" by splicing genetic material from a bean cell into a sunflower cell. Even though they could not regenerate the cells into plants, Secretary of Agriculture John Block called the step a breakthrough that "opens a new era in plant genetics."

German scientists have produced the "pomato," a plant that produces a potato underground and a tomato above. Several other studies are under way to breed a winter tomato that will be able to thrive in cold soil and a supertomato. Unfortunately, the supertomato may not taste better than the tomatoes we know and enjoy today. The idea is to produce a tomato that will be easier to grow, will be more disease-resistant, and will fit into cans better. We can hope that agrigeneticists will also find a way to make the supertomato even juicier and more flavorful than the varieties we now know.

Triticale, man's first artificially produced cereal, is being hailed by some experts as one of the grains of the future. Named for the Latin words for the wheat and rye genuses (*Triticum* and *Secale*), the cereal is highly nutritious and grows well even under difficult conditions.

Researchers at the University of Chicago have discovered a supersweet compound in a foot-high Mexican plant known as *Lippia dulcis*. The compound is about one thousand times sweeter than table sugar and could one day become the leader among artificial food sweeteners on the market.

Lipia dulcis, a plant from Mexico, may be the source of a new low-calorie sweetener.

TREE RESEARCH

Scientists are studying many aspects of tree improvement. For example, at the Morton Arboretum in suburban Chicago, botanists are transforming the silver maple into what they refer to as a more "civilized suburban tree." The scientists are crossing the silver maple with the red maple. The resultant hybrid combines the beauty of the red maple with the ruggedness and coping ability of the silver. Scientific study is also progressing toward producing a better box elder and a fruitless mulberry tree.

Meanwhile, in the upper Midwest, foresters have had some success in reducing the time that it takes for walnut trees to produce nuts. In

Vermont, breeders are trying to create maples that will drip sweeter sap. Quick-growing shade trees are another goal of researchers.

The main thrust in tree research, however, is in the development of trees that will resist disease. Blights and bugs have caused great harm to many trees, most notably the elm. Biologists are making great strides toward creating new trees that will grow faster and be more resistant to disease. They hope to be able to create exact replicas of fast-growing, disease-resistant species. The idea is to take a bit of tissue from a tree and, by treating it with chemical brews, cause it to sprout shoots and roots in the laboratory. Eventually, the clone could be planted in the soil. With conventional breeding, it takes about eight to twelve years to produce a seed-bearing tree. With cloning, a tree could be moved from laboratory to soil in one to two years.

Dr. George Ware carries a flat of elm saplings grown at Morton Arboretum near Lisle, Illinois. The new elms are resistant to Dutch elm disease.

One danger in cloning trees is that an insect or some unknown disease might wipe out entire plantations of similar trees. But foresters believe they can maintain genetic diversity in the woods to prevent such disasters, partly by breeding different trees for different sites.

WEED CONTROL

Helping to solve some of the problems farmers and plant breeders have with both established and new species of plants, scientists at the University of Illinois in Champaign have discovered a nonpolluting herbicide. It makes weeds literally commit suicide, while leaving crops such as corn and wheat unaffected.

The substance, an amino acid found in all plants and animals, is used by weeds in creating the chlorophyll that gives them their green color and utilizes sunlight to produce food. The amino acid is spread over crops at night, giving weeds a chance to load up on the light-sensitive substance. When the sun comes out the next day, the light reaction is so great, it kills the plants.

Discovery of the laserlike herbicide was a practical outgrowth of basic research into how plants produce chlorophyll.

An accessions clerk catalogs seed samples as they come into the National Seed Storage Laboratory in Fort Collins, Colorado.

The seeds are planted (left) in special paper toweling or in other materials for germination. Germination (right) may take from 6 days to 6 or 8 months, depending upon the requirements of the species.

MODERN PLANT RESEARCH

Plant research extends to plant life throughout the world. With the help of computers, government scientists at the Beltsville Agricultural Research Center in Beltsville, Maryland, are studying plant material (germ plasm) from all over the world. Germ plasm — the part of a cell that contains the hereditary information of a plant — has been used to breed new and better varieties of food crops. A new computer network has been created at Beltsville to allow plant breeders and researchers to plug into the data banks by phone and obtain whatever information they need in their genetic research.

In an effort to prevent the possibility that disease, natural disaster, or war might wipe out a specific strain of plant life or destroy a nation's or the world's food supply, billions of seeds are being stored in the National Seed Storage Laboratory in Fort Collins, Colorado. The seeds represent tens of thousands of plant species, assuring survival of the many species. The seeds are also available to scientists around the world for experimentation to continue the Green Revolution of plant genetics.

Checking germination tests

Evaluating germination tests

Beans in various stages of germination

Closeup of corn seedlings

Botanist checking for a sample in a cold-storage room

A research technician at work

SUPERANIMALS
3

Plant cloning, as was indicated in chapter 2, is already a reality and is getting more exciting every year. Animal cloning—though even more exciting—is a different story.

Genetic engineering in animals actually isn't new. It has been going on for centuries. Crossbreeding has created superior horses and cows and has even produced new breeds of dogs and cats. The turkeys and chickens we eat are important examples of genetically improved poultry. Yet, animal genetics and the cloning of animals are still mainly in the research stage.

One of the most obvious examples of genetic engineering in animals is to be seen in the horse. The first horses, thousands of years ago, stood only about twelve hands tall (120 centimeters, about 48 inches). Through breeding methods over centuries, horses grew taller and larger, into what today is called standardbred size.

Primarily just for the fun of it, less than a hundred years ago, some horse breeders began reversing the growth trend on purpose. Today there are many horse farms in this country where one can buy minihorses that are only thirteen inches tall. They cost thousands of dollars and are used mainly as pets.

The world has gained little from this example of genetic engineering of animals. But the prospects are vastly more exciting and useful for genetically changing or creating other animals that would be superior breeds. They would be healthier stock that could serve the world as increased and improved sources of food. Geneticists say they are close to creating the supercow referred to on page 13. Think of all the meat and milk such an animal could produce!

Left: Miniature horses are bred as pets. Hound Dog (top) is from the Elvis Presley collection. Sugar Plum (bottom) stands next to a Clydesdale.

FARM ANIMALS

Scientists are on the verge of creating super farm animals by injecting them with genes from other species, even from humans. Already, geneticists have doubled the size of mice by injecting them with human growth genes. They have hopes that pigs, sheep, cattle, and other livestock could be grown bigger and faster and would be more disease-resistant if similar techniques are used on them.

The supermouse was created in the laboratory by Ralph Binster of the University of Pennsylvania veterinary school in 1983. He fused the human growth-hormone gene with a mouse gene to produce a

The supermouse at the right has been injected with the human growth-hormone gene.

mouse that grew twice as big as normal size. It also passed on the growth trait to subsequent generations.

Federal researchers at the USDA lab in Beltsville, Maryland, are attempting to do the same thing with sheep and pigs. Ohio University scientists are implanting genes in mice and rabbits and plan to try pigs. Colorado State University is working with cattle.

Animals produced by these and other experiments would grow up faster on less feed, produce more meat and less fat, and suffer fewer illnesses, according to bioscientists.

EMBRYO TRANSPLANTS

Some pretty startling things have been going on in the animal kingdom lately because of a type of genetic engineering called embryo transplants. A quarter horse gave birth to a zebra at the Louisville Zoo in Kentucky. A donkey in a zoo in London, England, is expecting a zebra. A rare bongo was born to an eland, a common African antelope, at the Cincinnati Zoo in Ohio. The unusual births

This bongo antelope, born at the Cincinnati Zoo, was the result of the first transcontinental exotic animal embryo transfer. The embryo was one of a number removed from a single bongo at the Los Angeles Zoo.

are part of researchers' experiments to use abundant species to propagate endangered animals by means of embryo transfer.

In the embryo transfer process, fertilized eggs are retrieved from a donor animal and are placed in the receiving animal's womb. The first embryo transfer took place in 1898, when an embryo from one rabbit was shifted to another. In 1984, the first humans conceived in a donor and transferred to an infertile woman were born. More on so-called test-tube babies will be discussed in chapters 4 and 5.

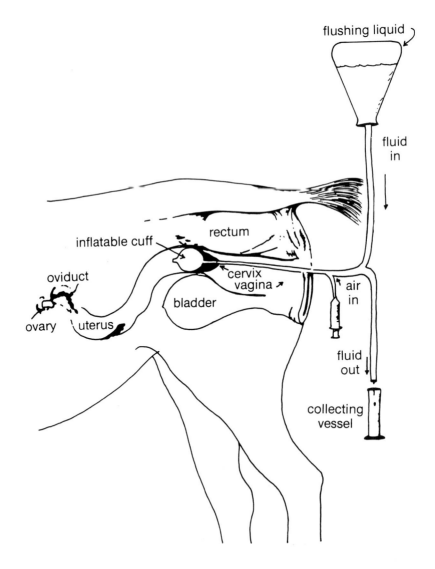

The diagram shows the process by which an embryo is flushed for transfer to a surrogate mother.

ENDANGERED SPECIES

Pollutants such as toxic chemical wastes, dangerous disease-fighting pesticides, and other man-caused environmental disasters have had serious adverse affects on both the plant and animal kingdoms in recent years.

"Man has created an environmental mess," says Dr. William Foster, head veterinarian at the Louisville Zoo. "Many animals have had their backs up against the wall as endangered species. We feel it is only fair to try to save animal species by using man-made techniques such as embryo transfer."

Foster says that with horses as substitute mothers, ten zebras can be produced in the surrogates in the eleven months it would take for a zebra mare to produce a single foal.

Embryo transplants may help to save endangered species. This baby zebra at the Louisville Zoo was born to a surrogate mother—a quarter-horse mare.

Other domestic animals could help preserve their wild relatives. At the Bronx Zoo in New York, a Holstein cow has given birth to a rare gaur, a blue-eyed bovine from the mountain forests of India.

Animal embryos can be frozen and flown to zoos in any part of the world where embryo transplants are being performed. The Cincinnati Zoo maintains a "frozen zoo," a kind of Noah's Ark on ice, with embryos and semen samples from dozens of species.

A BETTER HORSE

In Fort Collins, Colorado, a new breed of horse was sired in 1984 through the embryo transfer process. The horse was a bay filly appropriately named Eve who, having two mothers and a father, was the first of her kind.

Eve was born because her owner, Melanie Smith, loves horses and thought it would be wonderful if an American breed could be created specifically for jumping. The sport horse was born on the campus of Colorado State University as a result of experiments in which Eve had natural parents and help from a surrogate mother, Charlene. If all goes as Melanie Smith plans, one of Eve's offspring may be entered in the 1988 Olympics and win a gold medal for jumping.

Meanwhile, by splitting embryos, researchers have created identical twin horses. The twins, called Question and Answer, were born at Colorado State University's animal reproduction lab. Bioscientists flushed out a mare's uterus, then with the help of a microscope, found its dot-sized embryo. By means of a tiny tool called a micromanipulator, the cluster of embryo cells was divided. The halves, tucked in protective sacs, were then taken to be implanted in their surrogate mothers.

Thanks to embryo transfer, a prized female animal doesn't have to carry its offspring for the full term of pregnancy. The animal can become a four-legged factory for producing fertilized eggs, which are then implanted in other animals. In this way, a breeder can increase the number of offspring an animal can have over her lifetime and can simultaneously boost the animal's genetic contribution to the herd. Using embryo transfer in combination with artificial insemination, a

SPLITTING BOVINE EMBRYOS

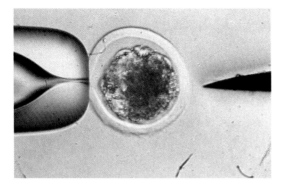

The flushed embryo is placed under a micromanipulator and grasped with suction through a pipette.

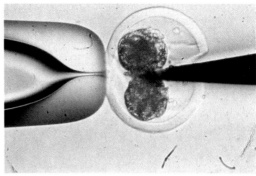

A microsurgical blade separates the cell mass into two sections.

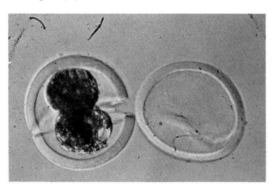

An empty shell *(zona pellucida)* is brought into proximity with the divided embryo.

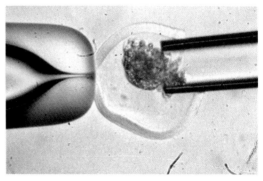

One of the sections is suctioned out of the original embryo and placed in the empty zona.

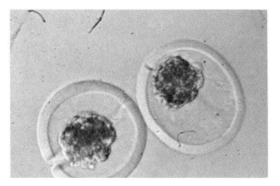

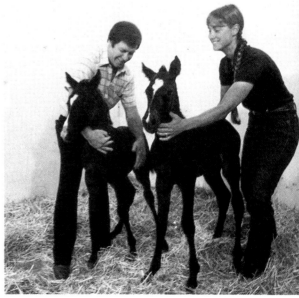

Above: The result is two embryos. In a short time the zonas reseal and each of the two parts can be treated as a single embryo.
Right: Question and Answer, twin colts born from a split embryo. They are genetically identical.

breeder can get hundreds of offspring, rather than six or seven, and improve his stock within a few years.

The primary benefit of twinning animals is that it will help researchers who are investigating the causes of infertility and other common reproductive problems in horses. Identical twins are ideal for such studies.

CLONING ANIMALS

Most animals, like most plants, reproduce sexually. A male and a female parent are required in order to produce an offspring. However, also as in some plants, some animals reproduce asexually; that is, without the help of both a male and a female parent. Some frogs, toads, salamanders, and other amphibians reproduce asexually by a method known as nuclear transplantation. Each offspring that results from asexual reproduction is a clone of its parent.

So far, the only animal that has been successfully cloned in a laboratory is the frog. Frogs are ideal laboratory specimens for genetic engineering study. They reproduce in great number, making them readily plentiful. Perhaps more importantly, a frog embryo matures into an organism with eyes, brain, liver, and other organs, much as a human infant does.

Another advantage in studying frogs is that the reproductive cycle from fertilization to the embryonic stage can easily be watched in glass dishes in the laboratory. One of the first scientific experiments you yourself may have observed in school was the cultivation of tadpoles by placing them in a dish of water with food. The in vitro (literally, "in glass"—test tube) method of studying a frog's development in the laboratory allows geneticists to have front-row seats to the whole creation process.

The idea of producing genetically identical animals or humans has fascinated the scientific world for years. Nature has its own method of cloning some animals. It achieves this by simply having a human or animal give birth to identical twins.

Identical twins—that is, genetically identical individuals—are not uncommon among humans, primates, cattle, and sheep; they are,

These identical twin calves were produced at Colorado State University from a split embryo. Notice their identical blazes.

however, rare in other species. Identical twins are born because a single cell splits and allows two separate embryos to form. This type of cell is known as a zygote.

Highly inbred lines of some mammals may already be considered genetically identical.

METHODS OF CLONING ANIMALS

One laboratory method for producing clones includes dividing early embryos. Another method involves inserting the nucleus of one cell into another. Researchers have found that in certain amphibians, nuclear transplantation from a body cell of an embryo into a zygote can lead to the development of a sexually mature frog.

Probably the best method now known for making genetic copies of any type of adult mammal involves inserting into an ovum the nucleus of a somatic cell—a body cell other than a germ cell—from that adult animal. But achieving this genetic miracle is expected to take years, if it is possible at all. The problem is that there is some evidence that most adult body cells of mammals are irreversibly differentiated. If this is true, it means that most body cells are so specifically programmed for their particular roles in the body that they cannot be changed.

Cell fusion is another method of cloning. In this method, two mature ova are fused, or one ovum is fertilized with another. Combining ova from the same animal is called selfing. Cell fusion may prove useful for transferring genetic material from a somatic cell into a fertilized single-cell embryo for the purpose of cloning. Selfing could result in pure genetic or inbred lines for use as breeding stocks.

Another possible cloning technique being studied involves producing chimeras. In genetics, a chimera is an organism composed of two or more genetically distinct tissues, or an organism that is partly male and partly female. The production of chimeras requires either the fusion of two or more early embryos or the addition of extra cells to

A pipette like this one is used to remove a single cell from an animal to produce offspring that are identical to the donor in every way.

an early stage in the development of mammalian embryos, known as a blastocyst. The cells added to blastocysts may be from different but closely related species.

Live chimeras between two species of mouse have been produced in the laboratory. But geneticists say that practical application of chimera technology to larger animals such as cows is not obvious at this stage of development. The main objective of chimera research today is to provide a genetic tool for a better understanding of embryo and fetal development and of maternal-fetal interactions.

Before scientists can successfully clone mammals, a better understanding of animal genetics is necessary. Before genes can be altered, they must be identified. This work has begun only in recent years.

EARLY ATTEMPTS TO CLONE ANIMALS

Cloning animals began in 1952 when two American scientists, Robert W. Briggs and Thomas J. King, working at the Institute for Cancer Research in Philadelphia, pioneered the nuclear-transfer technique for cloning. This method is now considered to be the standard approach to cloning animals.

In nuclear transfer, a nucleus is removed from a donor cell. It is then introduced into an egg obtained from a female of the same species. The egg's own nucleus is either removed or inactivated. This procedure allows the inserted nucleus to become the force that directs the development of the organism.

Briggs and King used frogs (actually tadpoles, not adult animals) in their experiments. They were able to transplant nuclei from frog embryos into egg cells from which the nuclei had been removed. These experiments produced some tadpoles.

But the scientists found that the older the transplanted nucleus was, the more likely their experiment was to fail. All the successful nuclei were obtained from embryos at an early stage of development, when the embryo was only a clump of cells. Briggs and King believe that the older cells had differentiated too far to express all the genes needed for the successful development of a new organism.

Some scientists believe that this crucial step in cloning—the need to work with nuclei from embryos in an early stage of development—may keep them from successfully cloning adult animals or mammals, including human beings. Others, however, are more optimistic. Some successes have already been reported, as we will see shortly.

About ten years after the tadpole clones were achieved, a British biologist, John B. Gurdon, successfully produced healthy adult clones of the African clawed frog. This species is often used in Europe as a laboratory animal. Gurdon took the nuclei from the intestines of young tadpoles, from cells that already were differentiated.

In 1975, Gurdon took the cloning of animals a large step forward when he produced tadpoles from the skin cells of adult frogs. Only one of the tadpoles grew into an adult frog. But Gurdon's work demonstrated that genes in fully differentiated adult cells could be made to express themselves after they had been "turned off."

OBSTACLES TO CLONING MAMMALS

A mammal is any animal that has a spine and that nourishes its young with milk from the mother's body. Cloning mammals presents additional obstacles for genetic researchers, though progress has been made toward overcoming them. One achievement has been the successful implantation of mammal eggs of several species into substitute mothers. The surrogate mothers have given birth to mice, cows, and even a baboon—animals that were not of the surrogate mothers' own natural species.

Pioneer work in cloning mammals was accomplished by a former student of Professor Gurdon. J. Derek Bromhall, a professor at Oxford University in England, attempted to clone rabbits in 1975. Instead of performing microsurgery on rabbit eggs, he experimented with chemical ways to take out the egg nucleus and introduce the transplant nucleus into the egg. This technique, called chemical fusion, worked better than microsurgery. Bromhall obtained four rabbit embryos from nuclear transplants.

At the same time, Dr. Clement L. Markert, a Yale University biologist, was attempting to clone a mouse. He had worked without

Another milestone in embryo transfer and surrogate mothering, this rare gaur is nuzzled by his Holstein mother.

success for some time with the nuclear-transplant method. Then he tried another approach.

Markert knew that parthenogenesis—the development of an egg without fertilization—is easily achieved in mouse eggs by chemical or mechanical means. A simple pin prick can start the division process, though this method has so far failed to allow the mouse embryos to develop into adult mice.

Markert experimented to see if the temporary introduction of the sperm into an egg might somehow promote complete development. To his disappointment, the eggs still did not develop any better than before. One reason the eggs failed to develop was that they did not have enough chromosomes. A normal fertilized egg, whether of a mouse, frog, or human being, has two sets of chromosomes—one from each parent. But what Markert had was a pronucleus, which has only one set.

A solution to the problem came from a British scientist's discovery of a substance known as cytochalasin B, which is produced by fungi. It inhibits cell division but does not interfere with the replication of chromosomes within the nucleus of the cell. When the egg, with a doubled set of chromosomes, is removed from the cytochalasin B, it continues to develop just like a normally fertilized egg.

The mice that resulted from the experiments were not, however, true clones. They did not have exactly the same complement of genes as the parents, inheriting only half the parents' genes.

Dr. Markert's method has a further serious drawback in that only female animals can be cloned by this technique. Male clones can be produced only by nuclear transfer, utilizing body cells containing two sets of chromosomes.

However, Dr. Markert is confident that his cloning method has an important future. He predicts that it will be widely used both to produce genetically identical laboratory animals for research and also to duplicate high-yielding dairy cows and other food-producing animals.

High-quality female animals produced by cloning could be bred to prize males as a way of speeding up selective breeding. Using other present methods of breeding, it takes about five years, or twenty generations, to produce purebred laboratory mice. It takes decades—even centuries—to produce a new variety of prize bull.

MILESTONES IN ANIMAL GENETIC ENGINEERING

In 1981, a scientific team, using rabbits and mice, claimed to have made the first successful gene transfer between animals. The history-making DNA experiment was carried out by a group of researchers, headed by Dr. Thomas Wagner of Ohio University, working on gene transfers in cattle.

That same year, scientists in Switzerland claimed to have achieved the first cloning of a mammal. Using cells from mouse embryos, they produced three mice that were said to be genetically identical to the original embryos. Each clone reportedly was produced by taking a nucleus obtained from a mouse embryo at an early stage of development and inserting it into a fertilized egg from another mouse. The original material in that egg was then extracted, leaving only the inserted nucleus.

After being cultured about four days, the egg was placed in the womb of a mouse that then gave birth to an offspring that was said to have all the genetic features of the embryo from which the nucleus had been taken. The offspring bore no relationship to the mouse whose egg had been used or to the mother that bore it. Two of the three mouse clones later produced seemingly normal offspring. The third died after seven weeks, but an autopsy revealed no abnormalities related to the cloning.

These colorful laboratory mice are multicolored for a purpose. Their colors identify the experiments in which they are being used.

One scientist, critical of the project, said that the experiment could not technically be called cloning. He went on to point out that although the three mice were identical to the embryos from which they were taken, they were not genetically identical to each other.

GENETICS ON THE FARM

Many genetic scientists are now convinced that they are on the verge of producing custom-made superanimals. They predict that the revolution in animal genetics will result in more productive farm animals to feed an exploding world population.

A farm worker poses with Maggie, a dairy cow, and her offspring. Maggie donated embryos, which were transplanted to surrogate mothers. This technique can increase herds much more rapidly.

Farmers have been applying the principles of genetics to livestock production for only the past thirty years, but results are already very impressive. Productivity has increased greatly just as a result of keeping accurate records on the desirable traits of breeding stock and through the use of artificial insemination. The U.S. dairy herd, for example, now numbers less than half what it was in 1950; yet it produces more milk.

Modern (left) and traditional (below) milking methods. The energy required for a cow to produce milk is dependent upon breeding and proper and sufficient feed.

Artificial insemination allows up to one hundred thousand cows to be impregnated each year from the sperm of one bull at a price of only ten to twenty dollars each. Artificial insemination is a billion-dollar business that now accounts for 100 percent of all commercial turkey breeding and for nearly two-thirds of all dairy breeding. Scientists anticipate making even greater achievements with such advanced genetic techniques as embryo transfer, sex control, cloning, and the ability to manipulate the genes of animals directly.

Breeders may soon be able to predict the sex of animals before they are born. A Colorado genetic engineering company reports it has developed the first reliable method of discovering the sex of cattle embryos. That is important to breeders of both beef and dairy cattle. A customer with a beef herd doesn't want a lot of cows, and a dairy farmer isn't interested in having a herd of bulls.

HEALTHIER ANIMALS

Scientists are using recombinant DNA, to create weapons against disease in animals. Cornell University researchers have boosted milk production in cows with a bovine growth hormone produced by stitching specific genes into cells. Similar hormones could increase wool production in sheep and meat yields in pigs and cattle.

Other researchers recently produced a safe, effective vaccine to protect cattle and other livestock from strains of hoof-and-mouth disease, one of the most serious and widespread diseases of animals. The U.S. Department of Agriculture calls the vaccine "an important breakthrough in animal genetic engineering" and the first production through gene-splicing of an effective vaccine against any disease in animals or humans.

Also through gene-splicing, researchers have developed a vaccine against scours, a fatal diarrheal disease of calves and young pigs.

Advances through genetic engineering in animals are coming regularly and rapidly now, though most of the current gene work is still only in early experimental stages. Even the scientists involved with genetic engineering in animals say that it boggles the mind to think what dairy cows will look like in twenty-five years.

Hogs have already changed because of genetic research. Thanks to genetic engineering, hogs are no longer being raised for lard but are being specially bred to yield quality lean meat. During the past thirty years, pork has become 50 percent leaner because of genetic research, formulated feed, and improved breeding practices.

SUPERCHICKENS

A recent breakthrough in gene-splicing by a California biotechnology firm may have a major impact on the economic benefits of the nation's $4.5 billion broiler-chicken industry and on lovers of fried chicken everywhere. The firm recently reported the first successful cloning of natural chicken growth hormones. The chicken growth is accomplished by taking genes that make the hormone from a chicken's pituitary gland and inserting them into bacteria. The genes then produce the hormone inside the bacteria which, in effect, become little hormone factories.

Researchers believe the discovery will help the broiler-chicken industry produce more birds faster and cheaper. In 1983, the industry produced nearly four billion broilers—enough to provide nearly fifty pounds of meat per person! Supplementary use of chicken growth hormones should help farmers bring flocks to market sooner, with lower feed costs and superior meat quality.

A NEW KIND OF APE

The chance mating of two species of ape in the Grant Park Zoo in Atlanta, Georgia, a few years ago produced a healthy offspring, the first reported ape hybrid. The birth gave support to a new theory of evolution. Researchers say the birth provides evidence for a recently advanced theory that new species may sometimes emerge through genetic juggling over a few generations, rather than from a prolonged series of small mutations over thousands or millions of years.

Shawn-Shawn, the hybrid ape whose mother is a siamang and whose father is a gibbon

The female hybrid ape, named Shawn-Shawn by genetics students, has a siamang mother and a gibbon father. The parents are farther apart genetically than are humans and great apes. Dr. David A. Shafer and Dr. Richard H. Myers, involved in genetic analysis of Shawn-Shawn and her parents, termed the hybrid a siabon.

"The fact that they could mate is significant scientifically because the chromosomes or genetic material of each parent are very dissimilar," Shafer reported. Gibbons have forty-four chromosomes, and siamangs have fifty. Shawn-Shawn has forty-seven chromosomes. Like most hybrids, she is believed to be sterile.

Three months after Shawn-Shawn was born, her mother rejected her for unknown reasons. She was taken to Georgia State University and cared for by graduate psychology students until she could survive on her own. At this writing, she is six years old.

MAMMOTHS MAY LIVE AGAIN

Some of the most unusual genetic engineering in animals today is being conducted in attempts to raise long-extinct creatures from the dead. Soviet scientists are trying to isolate living cells from a frozen mammoth found in Siberia. They hope eventually to bring the beasts back after ten thousand years of extinction. A spokesman for the Soviet Academy of Sciences has said that if mammoth cells can be cultured in the laboratory, "the final stage of the experiment—creation of a living specimen of the prehistoric northern elephant—would be completely feasible."

If the Soviet scientists manage to isolate living mammoth cells and culture them, the next step will be to combine one with a sex cell of an ordinary elephant. The nucleus of the elephant cell could be destroyed by radiation and be replaced with the mammoth's genetic material. Then the combined cell would be implanted into a female elephant. After about twenty months, the world's first artificially created mammoth baby would be born.

Meanwhile, California geneticists are working on a similar project, this one involving an ancient insect. A forty-million-year-old fungus gnat was recently discovered, set in amber, and studied. Insect

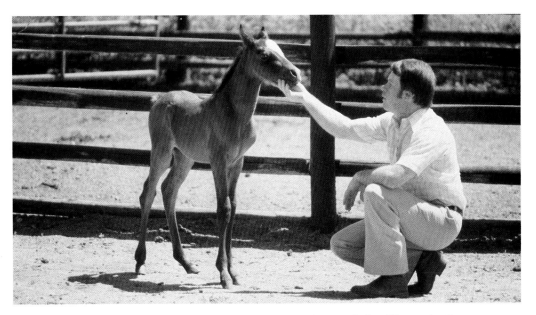

Dr. Ed Squires, director of the Equine Research Center, Animal Reproduction Laboratory, CSU, with Frosty, the first foal from a frozen embryo

pathologists George Poinar, Jr., and Roberta Hess of the University of California at Berkeley say the gnat exhibits the best-preserved fossilized soft tissue found so far. Its nucleus and other small cell structures were still intact.

Researchers hope to extract DNA from the gnat's cells. If that can be done, they will be able to compare the genetic makeup of the ancient fly with that of its modern counterpart. If intact DNA strands are rescued from the fungus gnat and implanted into the fertilized egg of a living fly, a gnat identical to the ancient one could result.

An alternative to cloning the gnat would be back-breeding. In the 1930s, German zoologists re-created the aurochs, an ancestor of cattle. They theorized that although this wild ox died out three hundred years ago, its genes should be present in cattle today. The scientists brought together cattle with traits similar to the extinct animal's, such as a heavy coat and long horns. Using these cattle, the German zoologists bred a "living picture" of the aurochs. Experiments are under way in this country to repeat the aurochs achievement.

In 1984, fragments of an extinct animal's genes were isolated from its preserved skin and reproduced in the laboratory for the first time. Russell Higuchi, a geneticist at the University of California at Berkeley, isolated genes from muscle tissue that was found in a stuffed quagga at the American Society of Biological Chemistry in St. Louis, Missouri.

Quaggas—tan-colored animals striped in front like zebras—once covered the central plains and veld in South Africa. They were killed off by man for their meat and hide. The last quagga died in 1883 at the Amsterdam Zoo in The Netherlands.

Even though the quagga's genes have been isolated, it does not mean scientists will be able to bring the extinct animal back to life. The isolated quagga genes are scattered into twenty-five thousand different fragments, each of which is being reproduced in a separate culture of bacteria. To recreate a quagga, scientists would have to reassemble all those gene fragments in the correct order. Then they would have to turn them on and off in the proper sequence to produce a baby quagga. The chances of that being achieved are, unfortunately, extremely remote.

STATE OF THE ART IN CLONING ANIMALS

How far are scientists from achieving a major breakthrough in the cloning of animals?

"By the indirect method I introduced several years ago, there is no theoretical reason why we can't do it now," says Dr. Markert, the Yale biologist who is regarded as one of the leading authorities on animal genetics. "It's more a matter of engineering development than basic research to make the techniques economically practical. That could take a few years of research, depending on how much effort is put in.

"If a big effort were made to finance the work, we would be able to clone mammals successfully in two to three years. The government is funding some research now. The cattle industry is putting some money into research, and once someone succeeds in cloning cattle, that would certainly convince private industry to invest in a large-scale effort to make cattle-cloning economically profitable."

Markert predicts that, unless scientists in other countries beat them to it, American geneticists could quite soon realize the spectacular achievement of cloning animals. In a recent report to the scientific community on the progress toward cloning animals, Dr. Markert gave this summary:

"Today, cloning other than by the production of identical twins is not possible. But in the future there are three principal ways in which we may hope to capitalize on some form of vegetative reproduction of desired phenotypes [organisms that have the same hereditary characteristics and that may look alike but that may breed differently because of dominance].

"These are first, the growth and multiplication of embryos in vitro. Second, the transplantation of nuclei from adult cells after treatment to restore the genomes [complete sets of haploid chromosomes; that is, having the full number of germ-cell chromosomes] to the embryonic zygote state [consisting of two mature cells capable of fertilization]. And third, the development of sufficient understanding of the genetic control of meiosis [cell division for reproduction] and egg production. This would allow us to genetically get around the present meiotic requirements in egg production. We could produce diploid eggs [having a similar pair of chromosomes for each characteristic, except sex] directly from oogonia [cells from ovarian tissue] that will develop to term without being fertilized at all.

"Any one of these three procedures will lead to the true and economic cloning of animals. But it is obvious that considerable basic research is still required, before we will achieve any of these goals."

The "white" elk is a color phase of the tule elk, which always breeds true. Scientists are studying the elk in an effort to determine the reasons for its consistent breeding pattern.

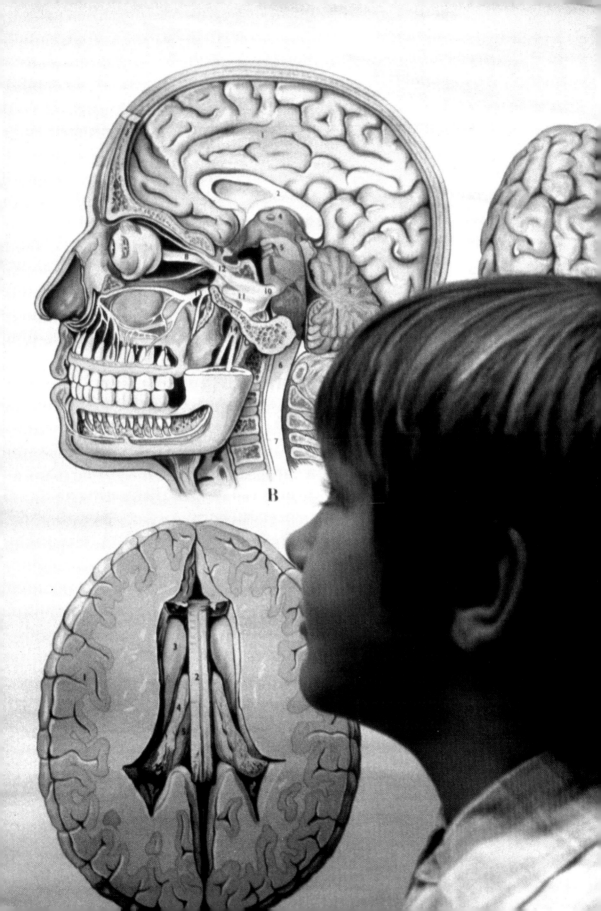

GENETICS IN HUMANS

4

Becky Summers was born a dwarf, a victim of one of about three thousand known genetic diseases. If not for very recent developments in human genetic engineering, she would have gone through life as a four-foot-ten-inch-tall adult. Now twelve, Becky, who was growing too slowly, gets regular injections of a growth hormone that was cloned in the laboratory from human genes. Last year, Becky grew five inches and is well on her way to enjoying a normal life as a more normally sized human being.

Mike Cassidy, fourteen, was stricken with melanoma a few years ago. Despite surgery, the deadly skin cancer spread throughout his body. Mike is praying that injections of genetically engineered interferon will pull him through. Interferon is part of the body's natural defense system for combating illness, another recent discovery of genetic research that may result in a cure for cancer.

Robby Newman, now twenty years old, has a genetic defect that affects every one of the hundred trillion cells in his pain-racked, palsied body. Robby's problem stems from a defect in the fertilized egg from which he was born. As a result, all of the cells in his body lack the instructions for making enough amounts of an essential enzyme that has a very long name—hypoxanthine guanine phosphoribosyltransferase (HPRT). The disease is called Lesch-Nyhan, a rare disorder that afflicts about two thousand Americans in varying degrees.

Robby lives strapped in a wheelchair during the day and in his bed at night, so his anguish and fear will not drive him to harm himself or take his own life. Despite the pain and frustration, Robby has kept his

Left: Perhaps one day this young student will make an important contribution to human genetics.

sense of humor. He also has kept his hopes alive that one day scientists may be able to cure him.

Over the past twenty years, drugs have been developed to help victims such as Robby so they now live into their twenties instead of dying in early childhood. Meanwhile, so-called gene doctors continue to work to find not only a better treatment for Lesch-Nyhan, but also a complete cure.

Geneticists also are working on more than sixteen hundred other diseases known to be caused by defects in single cells. These include muscular dystrophy, cystic fibrosis, hemophilia, several types of arthritis, and even hypercholesterolemia, an inherited trait that leaves one in five hundred Americans prone to early heart attacks.

Dr. Martin J. Cline of the UCLA Medical Center pioneered human genetic transplants. He planted a new gene in the living cells of two women to try to cure a fatal inherited blood disease.

There already have been some genetic engineering successes bordering on medical miracles. Researchers most recently cloned the antihemophilic factor, the substance that helps blood to clot and that those suffering from the genetic disease of hemophilia lack, making them susceptible to bleeding to death from even the most minor cut.

Other natural drugs that may soon be genetically engineered include the following:

△ **endorphin** — a natural morphinelike compound that controls pain and mood

△ **interleukin** — a protein that regulates the body's immune system from disease

△ **human serum albumin** — another protein available from blood that is used to replace blood lost from surgery or injury

△ **streptokinase** and **urokinase** — two enzymes that work within the blood system. They dissolve blood clots in the body, especially in the heart, brain, and lungs.

△ **tissue plasminogen activator** — a protein that dissolves blood clots. It works at the place of the blood clot without affecting the body's blood supply.

A SHORT HISTORY OF HUMAN GENE THERAPY

In 1869, an Austrian monk, Gregor Mendel, experimenting with pea plants, was the first to describe the mechanism of heredity.

Gregor Mendel (1822-1884), the Austrian botanist who laid the foundation for the science of genetics. Mendel's experiments led to his discovery of the basic principles of heredity.

In 1944, working with a bacterium that causes pneumonia, Dr. O.T. Avery of Rockefeller University discovered that genes are made of DNA. The double-helix structure of DNA was discovered in 1953 by Drs. Francis H.C. Crick and James D. Watson, pioneers in molecular biology.

The genetic code was cracked in 1965, when ribonucleic acid (RNA) was used to achieve protein synthesis in a test tube.

In 1970, two Johns Hopkins University scientists, Hamilton Smith and Daniel Nathans, discovered and used a new class of restriction enzymes, which are the chemical "scissors" that slice and separate DNA molecules.

In 1972, Dr. Paul Berg and colleagues at Stanford University combined the DNA from two viruses to perform a technique that yields what is called recombinant DNA. Three years later, recombinant DNA was inserted into host bacteria that cloned the foreign DNA. This work, by Drs. Stanley Cohen of Stanford and Herbert Boyer of the University of California at San Francisco, started the age of genetic engineering.

Two basic representations of DNA. *A, T, C,* and *G* stand for the compounds that are the basic structural units of DNA. The As in one chain are always linked to the Ts in the other and the Gs in one to the Cs in the other.

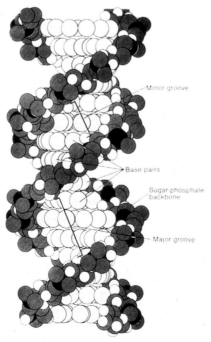
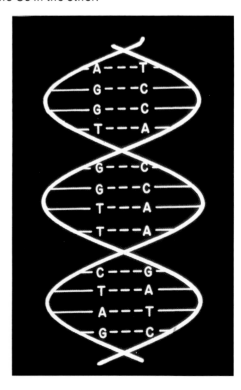

DNA

In 1978, sickle-cell anemia was diagnosed in an infant before birth by analysis of its DNA.

The year 1982 was a very big year in human genetic research. Human insulin was produced by recombinant DNA techniques. These techniques were first used for the prenatal detection of sickle-cell disease. The use of enzymes that specifically cut DNA, in combination with experiments that detect specific sequences of DNA, led to the development of a method for determining the location of genes. The technique has great promise for both promoting greater understanding of human genetics and assisting in the diagnosis of hereditary diseases.

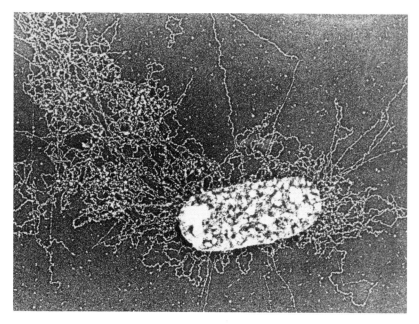

This photo, which looks like something from a science fiction movie, is actually a burst cell with DNA running out of it.

Also in 1982, the gene for rat-growth hormone was introduced into mice. It was discovered that adding zinc to the rat's diet caused it to grow to twice normal size, with far-reaching potential applications in humans. Not only will the hormone that triggers growth in humans allow thousands of children with growth disorders to develop normally; it may also help in the treatment of burns, bone fractures, and diseases of the elderly.

RECENT GENETIC DISCOVERIES

In 1984, biologists studying the molecular basis of organic life uncovered the basic roles played by specific molecular structure and composition in two quite different lines of research. American and Swiss molecular biologists discovered a DNA fragment—a specific sequence of genetic instructions—that appears to be common to such different organisms as yeast and man.

Meanwhile, a research team at the Massachusetts Institute of Technology was able to unravel the structure of molecules on the surface of the so-called T-cells, cells that counteract viruses, bacteria, or other alien cells that may invade an animal's or a human's body. The structural arrangement of these molecules enables T-cells to carry out their defense. The discovery sheds new light on the mystery of how the immune system works.

As the T-cell mechanisms are better understood, immunologists may be able to improve bodily defense against malignant cells. They also may be able to learn how to control T-cell activity to aid the acceptance of surgically transplanted organs, which T-cells now may attack as foreign.

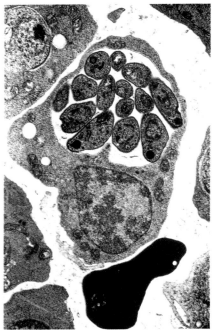

Left: Malaria is seen infecting the blood. This is one of the diseases that the human immune system does not completely conquer once the infection takes hold.

Below: Recombinant DNA research on insulin is done in the Eli Lilly laboratory. Insulin is used in the treatment of diabetes.

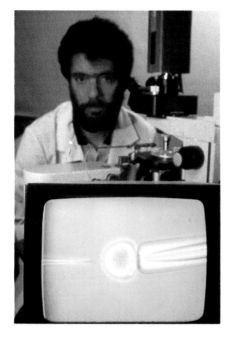
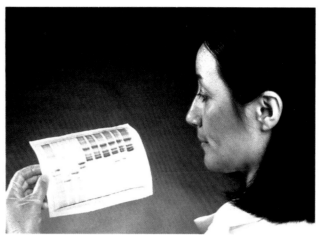

Left: A researcher uses special probes and a microscope to study genetic cell material. *Below:* Research into the human system involves the study of the genetic code. Here proteins are being analyzed.

The ability of microbiologists to isolate, analyze, rearrange, and synthesize the basic molecules of organic life is becoming a powerful tool in helping to discover the ways in which some of the most subtle mysteries of life function.

BIOLOGICAL SYSTEM BREAKTHROUGH

Genetic research late in 1984 helped scientists gain what they call a significant insight into the intricacies of the human biological system. Such insight could result in major medical benefits.

Geneticists at the National Jewish Hospital and Research Center in Denver, Colorado, reported the results of a twenty-year study to isolate one of the principal genes said to control the human immune system. The gene provides the blueprint for the key immune-system protein called a T-cell receptor. Scientists say this protein controls the action of a group of white blood cells whose role is to rid the body of infected, malignant, or foreign tissues.

"The finding will open the floodgates for basic research in a number of new but previously inaccessible areas," predicts John W. Kappler, a director of the Denver team of scientists. "It will enable researchers to produce large amounts of human T-cell receptor protein in the laboratory. It will provide scientists with a clearer understanding of how the immune system battles infections."

The discovery will help scientists in their quest to learn how the body's powerful defense weapons can be harnessed to fight disease. It was made possible by gene sequencing, a ten-year-old process by which the structure of a specific gene is charted. Because of the discovery many scientists believe they are now within twenty years of a complete understanding of the human central nervous system.

A CURE FOR AIDS?

The lives of thousands of people all over the world are endangered by the often fatal new disease known as acquired immune deficiency syndrome (AIDS). The disease slowly breaks down the body's natural immune system so it cannot fight back when attacked by viruses or other diseases.

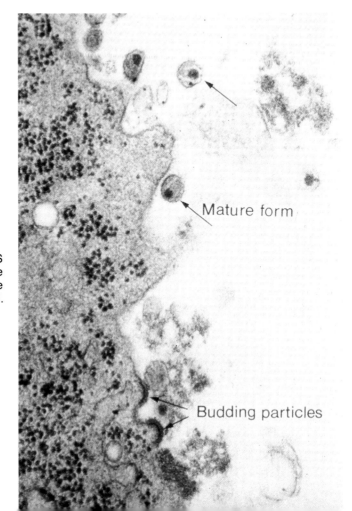

This photo shows the AIDS virus (HTLV-III). Both the budding particles and the mature forms are labeled.

At first, it was believed that AIDS was transmitted only by homosexual men. Then another theory was introduced, namely, that the disease came from improperly tested blood supplies obtained from donors in Africa. In 1984, researchers in France determined that AIDS is apparently caused by a type of human cancer virus. An American team identified the virus as T-cell lymphotropic virus, type III (HTLV-III).

Meanwhile, a cure is desperately being sought. Late in 1984, scientists in northern California took a giant step toward producing an experimental vaccine against AIDS. They accomplished this by cloning all of the genetic material from a virus believed to be the probable cause of the disease. Cloning the genes of the virus not only will help scientists to duplicate the virus in quantity for further study; it will also help to break the virus down into its component parts for research that could lead to a vaccine to fight AIDS.

HUNTINGTON'S CHOREA

Huntington's chorea is a disease that punishes its victims cruelly before it kills. It deprives them of control of their bodies and the use of their minds. In the United States, there are about twenty thousand victims of the disease and another hundred thousand who may have the gene that causes it. Many potential victims think it is better to live with the hope that they have not inherited the gene that causes the disease rather than to take the chance of finding out they have it.

Woody Guthrie, a well-known folksinger and composer of his time, died of Huntington's chorea. Many of Guthrie's songs have become American classics.

Among the victims of the disease was Woody Guthrie, a well-known folksinger who died in 1967.

In just a few years, scientists expect there to be a test for the Huntington's chorea gene. Researchers have already found a "marker" for the gene. It is a section of DNA that is on the same chromosome as the gene and seems to be inherited with it. By testing for the presence of the marker, doctors will be able to determine whether the gene is there, too.

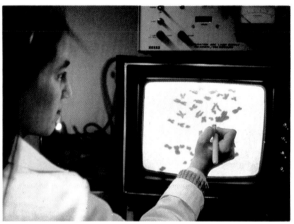

A technician uses a microscope and a video camera in counting chromosomes.

A researcher does chromosome studies in a child development and rehabilitation center.

The search for the Huntington's chorea gene was begun in 1979 by researchers at the Hereditary Disease Foundation in Santa Monica, California, and at the National Institute for Neurological and Communicative Disorders and Stroke (NINCDS) in Bethesda, Maryland.

Their discovery, which occurred about ten years before they had expected to find the marker, made genetic history. It was the first time that scientists had used recombinant DNA technology to find the approximate location of a gene without having any clue as to where it lay in the complex human genetic structure. Research was extensive, involving the study of forty-six chromosomes, about one hundred thousand genes, and a total of three billion base pairs—the rungs in the spiraling ladders of DNA.

Now the scientists expect to isolate the gene itself. Once they are able to study it, they hope to learn what it does and how it causes the vast range of neurological disturbances that are typical of Huntington's chorea. Eventually, their research could lead to a way to prevent, or at least treat, the disease. Their work may also lead to greater understanding of other inherited degenerative neurological disorders.

GENETIC KIDNEY DISEASE

Sometimes experiments in one field lead to failure, but their findings can be applied successfully to another. Such is the case with a drug that failed to work in therapy to control bodily seizures. It now is being used to help uncover the cause of genetic kidney diseases about which there is currently little understanding and for which there are no known cures.

The drug is a complex chemical compound called DPT, short for 2-amino-4, 5-diphenyl thiazole. Researchers in 1984 had reason to believe the drug could be good therapy for seizures and had given it

Unless a kidney transplant can be performed, this patient will have to undergo dialysis for the rest of her life.

to rats as a test. When the rats developed kidney disease, it became obvious that in its existing form, DPT could not be used in humans.

But one of the pathologists serving as a consultant on the research was not disappointed. Dr. Frank Carone, a professor of pathology at Northwestern University in Evanston, Illinois, saw the drug's potential in genetic kidney research. The rats' disease appeared to be similar to polycystic kidney disease (PKD), an incurable condition that affects about half a million Americans. It is a heredity disorder in which the tiny components of the kidney drainage system develop fluid-filled sacs called cysts. This causes the kidneys to swell up several times their normal size and eventually cease to function. The only way to prevent the patient's death is by kidney transplant or by artificial replacement of the kidney's cleansing function through a process known as dialysis.

Discovery of a compound that caused a condition in rats similar to PKD in humans made Carone see the possibilities of studying the disease in an animal model that could unlock the secrets behind the disease and lead to a cure. Scientists at Northwestern and three other medical centers are continuing their research with the help of a $1.4 million federal grant.

GENETIC RESEARCH INTO BLINDNESS

Doctors predicted early in 1985 that they may soon be able to determine with 90 percent accuracy whether women who carry a gene that causes blindness will pass the disease on to their sons. The prediction comes as a result of research by Dr. Richard Lewis, an ophthalmologist at Baylor University's College of Medicine in Houston, Texas, and Dr. Robert Nussbaum, a geneticist at the University of Pennsylvania. The researchers pinpointed the location of the gene that causes the disorder, choroideremia, in some males.

More than a thousand Americans are affected each year by choroideremia, which causes night blindness by the age of ten and total blindness as early as age thirty. It is one of several diseases that affect the retina, the part of the eye that receives light.

The defective gene is an abnormal X chromosome in females. It is so detrimental that the pigment in the back of the carrier's eyes appears speckled or mottled. Some carriers also may have mild night blindness and nearsightedness.

"Knowing the location of the choroideremia gene will allow us, within three to five years, to determine with more than 90 percent accuracy whether a child will be born with the disease," says Lewis. "We also will be closer to learning the cause of choroideremia and how to correct it."

HUMAN GENE IMPLANTS

The first authorized attempt to cure untreatable genetic diseases in humans by replacing defective genes with good ones probably will take place within three years, Dr. C. Thomas Caskey, geneticist at the Baylor College of Medicine Center for Human Genetics, predicted in mid-1984. "There have been breakthroughs in technology and in our understanding of genetic manipulation," said Caskey. "We still face many problems, but everything is now pointing in the right direction to make us optimistic that eventually it will work."

Key developments involve identifying the defective genes that cause specific diseases and making normal copies of these genes. Then scientists must develop techniques for getting the good genes inside cells where they can cure the diseases.

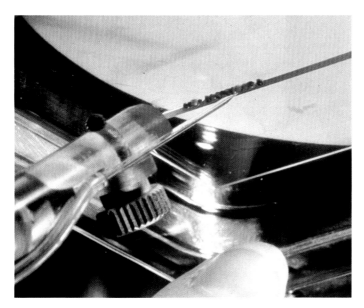

The genetic code is universal; therefore, a gene from a human cell will code for the production of the same protein in any other cell, regardless of whether or not that cell is human. Here messenger RNA from a human cell is microinjected into a frog egg. That egg will produce a specific donor protein.

One of the most recent advances is the development of a harmless virus that can be made to carry normal genes inside cells. This method has been used to correct genetic deficiencies in human and animal cells grown in laboratory cultures.

Caskey and his colleagues are planning to replace genes in patients suffering from immune deficiency disorders. These people lack the genes that program cells to develop immune defenses. Without immunity, victims usually die in childhood. The new gene therapy techniques probably will be effective only in cases where genetic diseases are caused by a single cell that is either missing or defective. Many other types of genetic disorders have complex causes that involve many genes.

Caskey says the most promising technique for treating genetic diseases appears to be the use of genetically engineered viruses that can carry repair genes inside cells with great ease. The carrier virus unloads its genetic cargo and then resides harmlessly in the cell without reproducing. The virus, called a retrovirus, was created by a team of scientists from the Massachusetts Institute of Technology. Animal experiments showed that the remarkable carrier virus was able to deliver cloned genes into stem cells. The genes became active, producing an enzyme, and were transmitted to all daughter cells.

In gene therapy, stem cells are removed from the bone marrow. Stem cells produce a variety of red and white blood cells and are involved in developing immunity. The stem cells are infected with viruses carrying the repair genes to correct a genetic disease such as the immune deficiency disorders or Lesch-Nyhan syndrome. The remaining bone marrow in a patient is destroyed, and the genetically repaired stem cells are replaced to repopulate the marrow with healthy cells, thereby curing the disease.

"The retrovirus system is simple and flexible," asserts Dr. Caskey, "and appears not to cause any harm."

GENETICS AND CANCER TREATMENT

Early in 1985, it was reported that the natural human cancer-destroying substance, tumor necrosis factor (TNF), had been obtained from a normal human gene by genetic engineering and was ready for large-scale trials as a cancer treatment. The use of the human gene encoding the TNF structure for production was the result of joint efforts by the Beckman Research Institute of the City of Hope in Duarte, California, and the Asahi Chemical Company of Tokyo, Japan.

"The recombinant TNF holds promise as a major advance in the treatment of many forms of cancer," said Dr. Charles W. Todd, chairman of the Beckman Research Institute's division of immunology. "Experiments show that TNF selectively attacks and destroys malignant cells with little or no effect on normal cells. It also is effective against various types of solid tumors in laboratory animals. This development opens the way for a major evaluation of the effect on many forms of human cancer."

Purified human TNF reportedly has shown lethal or growth-suppressing effects on a broad range of cancer cells, including some types of brain, nose and pharynx, lung, breast, stomach, and blood cancers.

Meanwhile, as research continues with TNF, genetic research in cancer continues on many other fronts. Another recent breakthrough involves the discovery of a new vaccine against feline leukemia. It was reported early in 1985 that the vaccine, described as the first effective anticancer vaccine, may represent a technological breakthrough in the fight against some human cancers and AIDS.

Dr. Robert Gallo, chief of the laboratory of tumor cell biology at the National Cancer Institute in Bethesda, Maryland, says that if the new vaccine for cats proves effective, "it could be an extremely important model for man." Gallo, who discovered the T-cell leukocyte in human leukemia and the AIDS agent HTLV-III, says the new cat vaccine "will certainly be the prototype for some of the things we would like to do in cancer research with people."

Feline leukemia is the leading cause of death among cats. Leukocell, the anticancer vaccine, was successful against leukemia in cats because researchers were able to isolate a protein called p15E from the feline leukemia virus. The protein interferes with the ability of a cat's immune system to respond to an injection of feline leukemia virus.

GENETIC COUNSELING

Discovering possible birth defects before a child a born is becoming a revolutionary new field in genetic research. Since the genetic formulas for several genes are known, scientists can hunt down and diagnose certain genetic abnormalities in the tissues of the human body. Doctors already screen for genetic defects before a baby is born, thanks to a diagnostic process called amniocentesis. In this procedure, the doctor removes from an expectant mother's womb a sample of amniotic fluid into which cells from the fetus have been shed. The cells are then inspected for possible genetic defects.

In the not-too-distant future, doctors will probably routinely screen adults to see if they carry faulty genes that they might pass on to their children. They also will be able to detect a person's susceptibility to certain diseases, such as hardening of the arteries, gout, and diabetes, even though the illnesses may not develop until years later.

Genetic counseling of pregnant women may help scientists to identify hereditary problems and find solutions before a child is born. Similarly, genetic counseling may also allow for screening people for faulty genes and correcting the problem before illness strikes.

Amniocentesis allows inspection of the fetus's cells for genetic defects.

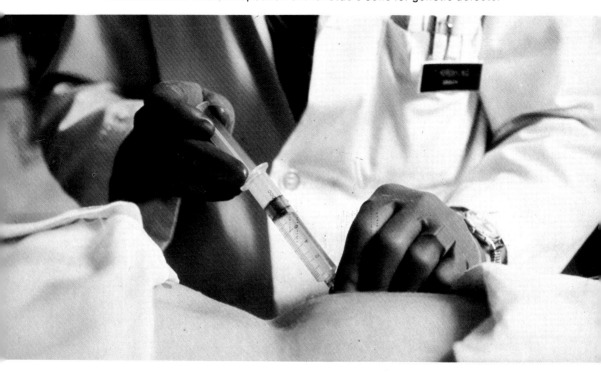

HUMAN GENETIC DATA BANK

Thousands of families, including generations not yet born, may benefit from the genetic information being stored at the world's first DNA bank, which has been established at Indiana University's medical school. Information learned from analysis of the human genetic material DNA that is being stored there will help doctors all over the country to diagnose genetic diseases. The data also will mean that more accurate genetic counseling will be available to families that are at risk of passing on diseases. The data bank will speed the research into identifying those characteristics of DNA that are indicators of various genetic diseases.

"By storing samples of DNA from critical relatives, the bank can be a way of preserving a family's past to better the future," says Nancy Wexler of the Hereditary Disease Foundation of America.

THE FUTURE IN HUMAN GENETIC RESEARCH

While there have been many significant advances in genetic research for humans, there have been setbacks as well. Just as encouraging reports were being circulated about the development of a growth hormone to help treat serious growth deficiencies in children and adolescents, the federal government in April 1985 called a halt to distribution of the hormone. Three patients who had received the substance died of a rare viral infection that may have resulted from the hormone's being contaminated with the virus. Researchers began work to positively determine the cause of the fatal virus. It is hoped that new batches of the growth hormone will be free of contamination for future use in this important new genetic breakthrough for victims of growth deficiencies.

Research in human genetics is now expanding rapidly. It is leading to increasingly precise ways of preventing or treating some of the two thousand or more inherited disorders that afflict human beings. At the same time, human genetic engineering has produced a wealth of new ideas and techniques that are laying the groundwork for a new medical science in the twenty-first century.

Above: An ovum is fertilized in vitro (in glass). *Below:* The first in vitro quadruplets are pictured with their sister.

TEST-TUBE BABIES
5

Under normal circumstances, a woman becomes pregnant when an ovum, or egg cell, released by her ovary during ovulation, is fertilized in the fallopian tube by a single male sperm cell that has traveled up from the vagina. After the fertilized egg undergoes a number of cell divisions, the tiny clump of cells enters the uterus. There it burrows into the wall and develops until birth.

That is how most women give birth to children. But there are nearly two and a half million couples in America alone who, for one reason or another, are infertile and cannot conceive a child. Most infertility is caused by missing or malfunctioning fallopian tubes, infections, defects in the male's sperm, or hormone imbalances that interfere with a woman's normal production and release of eggs.

Little or nothing could be done for these childless couples until only a few years ago. There has, of course, always been the alternative of adoption, but today there are other solutions to the problem—artificial insemination, in vitro fertilization, and embryo transfer. Artificial insemination is the process of inseminating a woman with a donor male's sperm. The sperm may be her mate's or, if necessary, another donor's. In vitro fertilization is carried out in the laboratory—outside the body. The egg is fertilized and then is implanted in the woman's body. In some cases, it may be a donated egg that is fertilized in vitro with the mate's sperm and then transferred into his wife. In embryo transfer, an embryo is transferred from a fertile female to one who is infertile.

THE BEGINNINGS OF THE TEST-TUBE BABY BOOM

In 1978, a British couple conceived a child by in vitro fertilization and made not only a baby, but also medical history. Gilbert Brown, a thirty-eight-year-old truck driver in Bristol, England, and his wife Leslie, thirty-one, had been married for nine years. Though they had tried for years, they were unable to bear children. Doctors said that the reason was a blockage of Mrs. Brown's fallopian tubes. They told her there was no chance she could ever conceive.

As a last resort, the Browns went to the District General Hospital in the old textile mill town of Oldham and spoke with two doctors about whom they had heard. Patrick Steptoe, a gynecologist at the hospital, and Robert Edwards, a physiologist at Cambridge University, had become known for their experiments with in vitro fertilization.

The doctors had good news for the Browns. Their experiments led them to believe they could, despite Mrs. Brown's fallopian blockage, help her give birth to a normal child. And so began what some say is the most sensational birth since Adam and Eve conceived.

The way the doctors would bring Mrs. Brown's baby into the world was this: A ripe egg would be removed from her ovary. It would be placed in a laboratory dish and sperm from her husband would be added. After incubating the ovum as it started to divide, the doctors would finally place the developing embryo in the uterus. There it would become implanted and continue to grow into a fetus in what would seem to be an entirely normal way.

Laboratory culture dishes are prepared to receive eggs for fertilization.

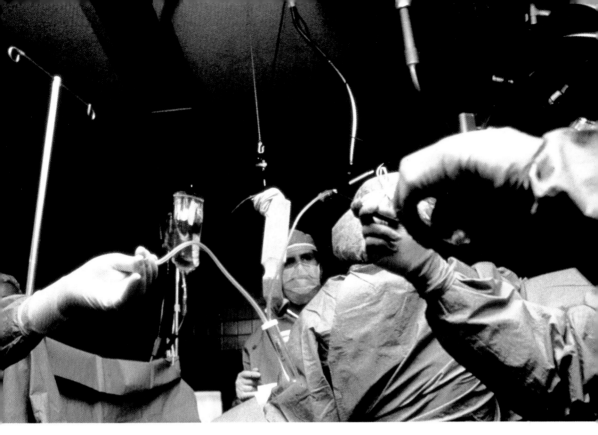

Fluid containing developing eggs is removed with a laparascope and transferred to a vial.

But the doctors admitted to the Browns that despite many previous attempts to transfer externally fertilized eggs, a successful, full-term pregnancy had not yet been achieved. Still, the Browns wanted their chance to have a child. They told the doctors to go ahead and use them as guinea pigs. They also made a $565,000 deal with the London *Daily Mail* to give the newspaper the exclusive story of the first baby ever to be conceived outside the womb.

In November 1977, Mrs. Brown was given hormonal injections to stimulate her egg cells to mature. Afterward, a small cut was made in her abdomen and several eggs were removed from her ovary. They were placed in a laboratory culture medium and exposed to Mr. Brown's sperm. At least one egg was fertilized. The resulting conceptus began dividing, first into two cells, then four, then eight.

A few days later, the conceptus reached the blastocyst stage, when a group of cells form a hollow sphere. (Under normal circumstances, fertilization and this initial division would take place as the egg traveled through the fallopian tube to the uterus.) At this point, the laboratory conceptus was introduced into Mrs. Brown's womb.

Sometime later, as the pregnancy progressed, Drs. Steptoe and Edwards made tests from which they said they could determine the sex of the child by examining chromosomes. But Mrs. Brown asked that they not tell her whether she was to give birth to a boy or a girl. "I've been waiting too long for this to be denied the surprise of learning whether the baby is a boy or girl at birth," she said.

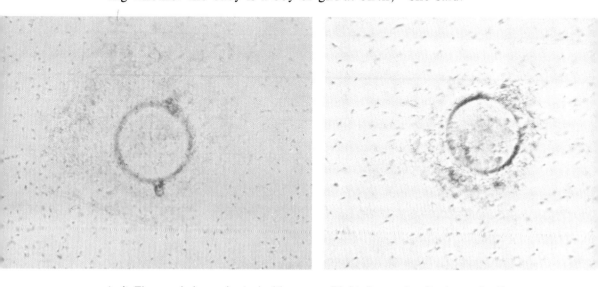

Left: The egg is inseminated with sperm. Right: Seconds after insemination, hundreds of sperm swarm around the egg.

Meanwhile, Mr. Brown became more than the typical nervous expectant father. "I didn't know we were to be the first parents to have a test-tube baby!" he later said in an interview. "If we were, I wish we weren't!"

As the blessed event neared, word of the amazing birth spread throughout the world. Reporters and television crews came from many foreign countries to cover the big story. Most, but not all people, thought it was wonderful. Some who did not like the idea picketed the hospital. One day, a bomb threat caused a sensation in the hospital, but no bomb was found. Then a rumor circulated that someone was going to kidnap Mrs. Brown. An almost circuslike atmosphere began to prevail.

Late in her pregnancy, Mrs. Brown underwent advanced testing. Ultrasonic scanning checked the position, size, and shape of the fetus as it developed. Hormone levels and fetal heartbeat were monitored. Anmiotic fluid was taken from the womb to be checked for signs of any genetic diseases.

Finally, the suspense was over. On July 25, 1978, at thirteen minutes before midnight, Mrs. Brown gave birth to a five-pound, twelve-ounce blond, blue-eyed daughter. The normal, healthy infant was named Louise Joy.

Despite all the publicity and high expectations, the birth caught many scientists by surprise. They had talked for years about fertilizing the human egg in a test-tube, but widespread doubt still existed that it could be done successfully. The achievement immediately drew criticism from many people, especially conservative churchmen. Steptoe tried to calm their fears that he was planning to create babies outside the womb. "All I'm interested in doing is helping women who are denied a baby because their tubes are incapable of doing their small part," he explained.

Controversy over test-tube birth methods continues to this day, while Louise Joy Brown at this writing is a happy, healthy seven-year-old living somewhere in England. Hundreds of other test-tube babies are also living normal, healthy lives in various parts of the world as a result of the pioneering work done by Drs. Steptoe and Edwards.

Steptoe and Edwards repeated their achievement in 1979 when the first test-tube boy, Alastair Montgomery, was born in Scotland.

Afterward, the doctors announced they were working on techniques of fertilizing eggs with the sperm from men who previously were unable to conceive because of low sperm counts. Other experiments were begun in hopes of producing the first test-tube twins.

In order to produce twins outside the womb, two eggs must be taken from a woman's ovary. Most women produce one egg a month, but a woman can be injected with Pergonal, a hormone that stimulates the follicle and encourages the development of many eggs.

Steptoe said the rest of the operation would be the same as for a single child. First one egg would be sucked up out of the woman's abdomen and transferred into a culture dish with nutrients and sperm cells, where it would be fertilized. Then the second egg would be extracted and the procedure repeated.

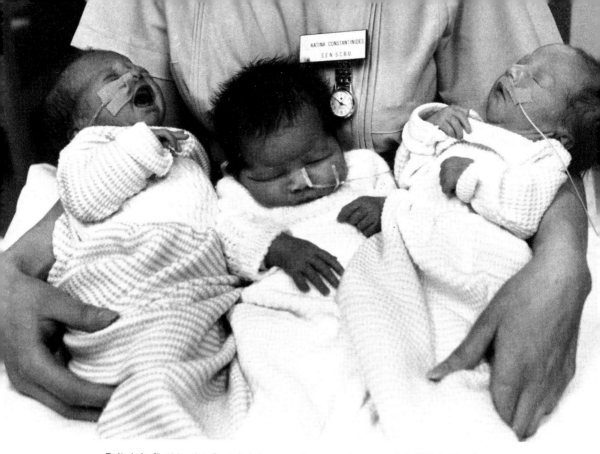

Britain's first test-tube triplets were born on January 21, 1984. The boys, flanking their sister, are the world's first in vitro identical "twins."

In two and a half days, after each of the fertilized eggs had undergone cell division, the two embryos would be transferred to the uterus. In nine months, the mother would give birth to fraternal, nonidentical twins.

TEST-TUBE BIRTHING TODAY

About 350 babies have been born worldwide so far by in vitro fertilization. More than seventy clinics in the United States now offer the technique to infertile couples at a cost of about four thousand dollars. At first, the pregnancy rate for the process was only 5 percent, but now it is 30 percent.

The most recent artificial fertilization development involves freezing embryos for later transfer. No one knows how long an embryo can be frozen and still be useful when it is thawed. In the process, life is put on hold in a state of suspended animation, frozen in liquid nitrogen at minus 196 degrees centigrade. Then it is thawed.

While in vitro births have drawn widespread criticism, freezing embryos has raised new legal questions. In 1974, an infertile couple, Mario and Elsa Rios, died in a plane crash. They left behind two frozen embryos in an Australian clinic. The couple's estate, valued at more than a million dollars, was to go to the dead man's son by a previous marriage. But what about the frozen embryos? Who owned them? Further complicating matters, Mr. Rios was not the biological father. The sperm of an anonymous donor had been used.

A committee in Australia recommended that the embryos be destroyed. But women all over the world offered to act as surrogate mothers to help give birth to the frozen embryos. The controversy never resulted in a birth.

Meanwhile, in late March, 1984, a New Zealand woman gave birth to a daughter, Zoe, who became the first baby ever produced from a frozen embryo. The five-and-half-pound girl was conceived by in vitro fertilization and born by Caesarean section.

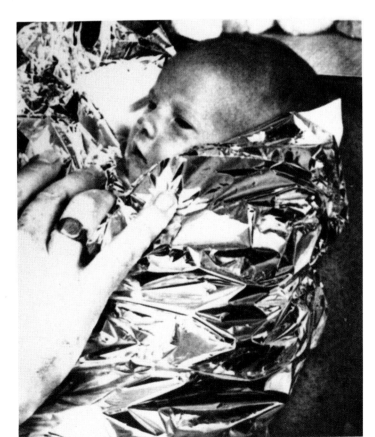

Zoe, the world's first authenticated frozen embryo baby, is shown at the age of two weeks. She was born in Melbourne, Australia.

The mother's ovum was fertilized in a laboratory with her husband's sperm. The embryo was then frozen, reportedly for two months, before being implanted in the woman's uterus. The parents' names have been kept secret so that the family can live in privacy.

TEST-TUBE BIRTHING PUT ON HOLD

America's first test-tube baby, Elizabeth Carr, was born on December 28, 1981. She is a happy, healthy preschooler today, like about a hundred other infants conceived outside their mother's wombs in this country since then. An estimated fifty thousand babies are born in the United States each year through artificial fertilization. Yet, a federal moratorium on research into in vitro fertilization exists in this country.

In 1975, the former Department of Health, Education, and Welfare said it would no longer fund any proposal for research on human embryos or on the external fertilization of human eggs unless that proposal was reviewed by its ethics board. The moratorium was never lifted, severely restricting further research on artificial birthing. Without federal research funds, American doctors have had to borrow information and techniques from scientists in England and Australia or rely on modest patient fees or private donations. The federal ban on research has resulted in a "cloud of ethical or moral suspicion" that hangs over artificial birthing, a cloud that doctors and scientists working in the field say is unwarranted.

IS TEST-TUBE BIRTHING RIGHT OR WRONG?

Not everyone wants test-tube babies. Some scientists, researchers, government officials, and many conservative churchmen and lay religious people are up in arms about what they call tampering with the normal or traditional method of human birth. Many criticize test-tube births as being immoral and against the will of God.

"What we are trying to do is help nature," Dr. Steptoe explains. "I cannot see that there is anything immoral about that. We are changing, testing, adapting all the time. With more teams working on this research, I believe we can achieve a success rate of 50 percent in in vitro births. There is no doubt we can do it. We are at the end of the beginning."

Some, however, believe we may be approaching the beginning of the end. "The issue is how far we play God, how far we are going to treat mankind as we would animal husbandry," argues a British member of Parliament, Leo Abse.

His sentiments are echoed by many elsewhere around the world who see test-tube fertilization not as a humanizing technique, but as a dehumanizing one. They fear that in vitro births take the moment of creation outside the human body and put it into a mechnical or laboratory environment.

A micrograph is an image of what the microscope sees. This micrograph shows human sperm.

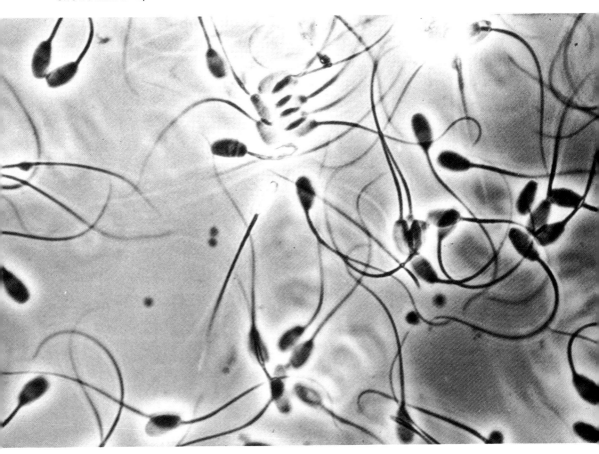

Critics say they accept the procedure for couples who want a child and cannot otherwise conceive. But they maintain that society is built around the family. They wonder what happens to the family once sex is divorced from procreation.

The Roman Catholic church in this country opposes test-tube birthing, just as it does other medical and scientific techniques for procreation and birth control. "Interference with nature is not acceptable in any form," one high official in the Catholic church states. "For that reason, the pope has condemned artificial insemination, even with the husband as donor."

A leading Protestant theologian, Paul Ramsey, maintains that test-tube births are immoral because of the uncertainties involved. "The parents' right to have children is never so absolute as to justify such induced risk to the child," he has said. "The rights of the child-to-be should be considered."

A Jewish theologian disagrees. "When nature does not permit conception, it is desirable to try to outwit nature," says Rabbi Seymour Siegel, a professor of ethics at the Jewish Theological Seminary in New York City. "The Talmud [the body of early Jewish religious and civil law] teaches that God desired man's cooperation."

In vitro fertilization is not illegal in the United States. Federal and state laws do not forbid it. Yet, the ongoing government freeze against research funding continues to slow development in the field and fans the flames of moral or ethical prejudice against it.

Many American couples facing childless marriages attended an ethics advisory committee hearing of the Department of Health, Education, and Welfare in 1979. One woman, Judy Jones, a twenty-seven-year-old housewife from Farmington Hills, Michigan, testified that she and her husband had tried for four years to conceive a child, but her fallopian tubes were damaged and had been removed.

"I've undergone two operations and spent a total of six months in the hospital," Mrs. Jones said, near to tears. "But nothing has worked. My doctors have told me we can never have a family. We've tried to adopt a child, but even after two years, haven't been successful. We desperately want a family. Please give us and people like us a

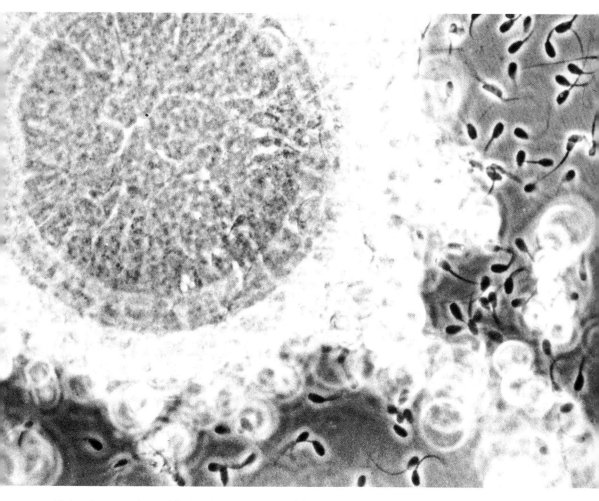

This micrograph enables us to see an egg with sperm swarming around it.

chance. I know it involves many moral and ethical issues, but in vitro birth may be our only hope."

Many of the critics of test-tube birthing have the same reservations about cloning. The controversy surrounding in vitro fertilization and human cloning can be traced to one common argument—the risks and morality of laboratory methods of conception. "Is it moderate-risk recombinant DNA research?" they ask. "Or is it playing God?"

CAN A HUMAN BE CLONED?
6

So you want to clone a human being. All right then, here's how. Ready, get set, go!

RECIPE TO CLONE A HUMAN

>1 egg from one female
>1 cell from a male or female
>1 womb from a female
>Assorted laboratory equipment

Remove egg from female's fallopian tube shortly after ovulation. Pluck out cell nucleus. Scrape from a male's or female's skin a single cell containing in its nucleus all the genetic instructions to create a new person. Transplant nucleus into enucleated egg. Cytoplasm will start the genetic code into action. After egg has divided several times, place in womb of a healthy female. Incubate about nine months. Yield: a genetic duplicate of the cell donor with the same sex, physical, and mental potential.

This technique for cloning a human is the one by which David Rorvik claimed in a book written in 1978, *In His Image,* that the first human clone had been born. The book caused a sensation in the

Identical twins such as these are nature's way of cloning. Both develop from the same egg and are genetically the same.

medical world, even as skeptics doubted that a human had been cloned. Rorvik finally admitted that his book was fiction, not fact.

But could a human be cloned that way? "The method of cloning a human described by Rorvik is conceivable, but not likely," Dr. Clement L. Markert, Yale biologist and authority on cloning, commented at the time of the controversy. "We clone mammals now by nuclear implantation, which is the technique Rorvik described. But his book is just science fiction.

"I believe Rorvik's claim that a human already has been cloned sold a lot of books and increased public interest in cloning. I'm not

The man at the right is David Rorvik, the author of *In His Image*, the book in which he claimed that a son had been cloned of a California millionaire.

sure whether that has helped the field of genetics or cloning. It's hard to say about that. It made a lot of people worried."

Dr. Markert and others are not sure it is essential or even necessary to clone humans. They can see the value in plant and animal cloning primarily to improve species and thereby increase the world's food supply. But they see no real gain from cloning human beings.

Most scientists do not discount the possibility of one day cloning a human. "I would like to say that it is impossible, but it is not," says an authority in reproductive physiology, who wishes to be anonymous. "All the bits and pieces—although developed in other animals—are already lying around. It might be possible to clone a human, if you had lots of people to use."

Although it may look easy on paper, cloning a human would be an extremely complex operation in practice. So far in the evolution of the art of cloning, after years of trying, only a few frogs and some female mice have been cloned. Even in the frog, no clone has been produced from an adult cell. Scientists say that adult cells are too specialized to permit them to be used to form new organisms.

Rorvik claimed that a doctor had somehow achieved a biochemical breakthrough that induced cells "to forget their specialties and start all over again." Many genetic authorities say this claim is absurd. "Anyone capable of making such a 'breakthrough' would have solved most of the problems of tissue regeneration and cancer development," argues Dr. J. Bromhall of Oxford University.

"It is possible that once a cell has fully specialized, it is irreversibly changed," says Dr. Markert. "If so, adult human cloning may never be possible."

Another leading scientist agrees. "It is highly unlikely that we are ever going to be able to clone human beings," says Dr. Bernard D. Davis, professor of bacterial physiology at Harvard Medical School. "Scientific considerations have reduced fears that humans can be cloned, and I say, thank God."

Davis further argues that to impose on cloned humans an "identity crisis of unprecedented dimensions" would be a terrible moral problem. While he sees great benefit to mankind if superior cattle and other livestock could be cloned, he opposes human cloning on moral grounds. Scientifically, he insists, it can't be done anyway. All body cells do not have the same gene content, he points out; therefore, a human being could not be created by cloning.

Nobel laureate James Watson, co-discoverer of DNA's double-helix structure, which helped open up the age of modern genetic engineering, agrees that human cloning is something for the future, if at all. "What's to be gained? A carbon copy of yourself? Oh, if the Shah of Iran wanted to spend his oil millions on cloning himself, that's fine with me. But if either of my young sons wanted to become a scientist, I would suggest he stay away from research in cloning humans. There's no future in it."

RECENT HUMAN CLONING RESEARCH

As controversy over cloning humans continues, researchers are apparently fitting together more pieces of the puzzle. In 1979, Dr. Landrum B. Shettles reported a new discovery. He said he had taken germ cells from women undergoing various gynecological tests and from men having testicular surgery. Eggs were withdrawn from the follicles of women who were operated on at about the time of their ovulation. The eggs were then incubated in follicular fluid.

After about three hours, it was reportedly possible to remove the outer layer from the eggs to detect polar bodies in them. These showed that an egg was ready for fertilization and that nuclei (genetic material) could be drawn from the eggs. The procedure left the eggs intact but without their nuclei.

Dr. Shettles suggested that the next step in cloning a human would be to replace the nuclei with genetic material from the person who was to be duplicated. To accomplish this, spermatogonia, precursors of sperm cells, would be taken from male volunteers. Spermatogonia would be used because they are less active than mature sperm cells.

The doctor said that he isolated nuclei from the spermatogonia and inserted them into the denucleated egg. Three sperm nuclei transfers were successful, he claimed. These resulted in fertilization of eggs and egg division to the multicelled blastocyst stage, the stage at which an egg fertilized in the usual way would leave a woman's fallopian tube and become implanted in the womb.

Dr. Shettles ended his experiments at this point, though he said there was every indication that each specimen was developing normally and could have been transferred into a womb. It would then, he maintained, develop into a human containing genetic material only from the male sperm donor.

Several scientists said that some questions about the experiments remained unanswered. Mature eggs can be obtained from women only after careful monitoring of their ovulatory cycle, and there was no evidence that that was accomplished.

Another objection was that a blastocyst would have to be analyzed in order to determine that each of its cells contained chromosomes matching those provided by the male donor. This would have to be done before it could be proved that the blastocyst was really the start of a human clone.

Other critics asked how a nucleus from a spermatogonia might have fertilized an egg without a nucleus and have thus led to egg cell division. During normal fertilization, a sperm cell enters an egg, its nucleus and the egg's nucleus merge, and chromosomes that are present in both nuclei merge. The egg then divides into two cells, four cells, and so on, with each cell containing genetic material from both the sperm and the egg.

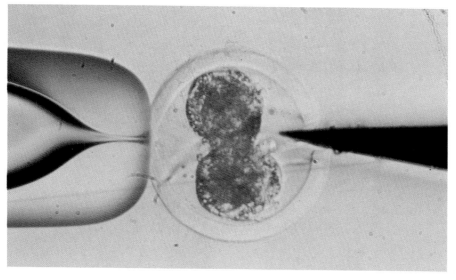

Although scientists are able to create animal twins by splitting embryos, they are not yet able to clone mammals. Cloning, especially of humans, is a highly controversial subject.

An interesting concept was revealed recently by another scientist, who says that by presently described cloning methods, it is impossible to make copies identical to an individual donor of cells being cloned. "It is a false notion, so far undisputed by scientists, that the product of a single successful nuclear transplantation is a 'clone,' " said Dr. Paul R. Gross, director of the Marine Biological Laboratory, Woods Hole, Massachusetts. "That is, an identical copy. There is no possibility that a literal copy of the donor individual can be produced by the insertion of a somatic nucleus into a recipient cytoplasm of a conveniently available egg. A roughly similar individual, yes. But a carbon copy, no.

"We have, therefore, a long way to go. It is, however, a way that should not be blocked nor impeded by hysterical appeals to 'Stop Xeroxing People!' "

A COPY OR A MUTANT?

If a human being is cloned, there are risks that it would not be a carbon copy of the original but a mutant, warn some scientists. However, chances for mutation in the clone depend on what kind of cell is used to fertilize the egg in the cloning birth. A sperm cell has a good chance of producing a direct replica, without mutation. But because it carries only twenty-three chromosomes, half of the normal total per cell, the offspring would be created from only half of that parent and would not be a direct genotype.

A CLONED PARTS BANK

Some scientists who are opposed to human cloning suggest that a cloned "parts bank" could be created. If the cortex could be prevented from developing in the clone's brain in the early stages of development, a genetic duplicate could be created. The clone would not be a person, but a parts bank.

The clone's heart, kidneys, and all vital organs could be transplanted back to the original person or used in others in need of them. The clone's parts would be far superior to other transplants because the recipient's body would not reject the new organ. Crippled people

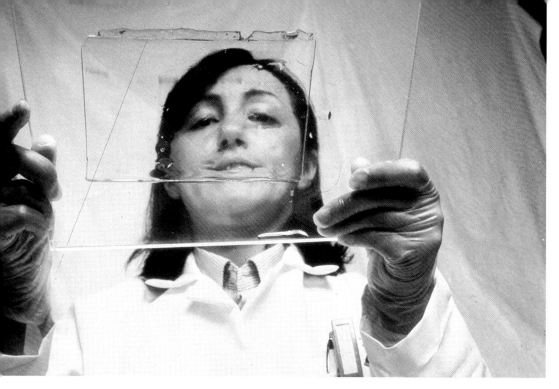

The search for genetic information continues and becomes increasingly sophisticated. Here a technician analyzes individual genes on DNA by a process called electrophoresis.

could get new limbs and cancer victims could get new organs. The potential benefits of a cloned parts bank are called amazing.

THE FUTURE OF CLONING HUMANS

Most scientists agree that not only will critics of human cloning have to change their minds about the possible miracle, but that also more medical and laboratory miracles will have to occur before the first human being is cloned.

At present, the most important breakthrough may involve working with adult human cells. This is the main stumbling block in the cloning of animals. Yale's Dr. Markert has said, "It may be that the nuclei from adult cells just simply cannot replace an egg nucleus. If that's true, then no technology is going to bring about cloning, in adult mammals or in human beings."

Can humans be cloned? The answer to that mighty question lies somewhere in the future. And the answers to how far into the future and whether or not it will ever really happen depend on which scientist you are listening to.

GENETIC ENGINEERING AND CLONING

7

GOOD OR EVIL?

Out of bits and pieces of human and animal bodies and bones, in a smog-shrouded laboratory high in the European mountains of Transylvania, a Swiss student of the occult sciences learned the secret of infusing life into inanimate matter. He created a live, but grotesque, misshapen being that ultimately destroyed him.

The laboratory creation of Dr. Victor Frankenstein had come to a British writer, Mary Woolstonecraft Shelley, in a dream. Her book about the man-made monster, which came to be called Frankenstein, caused a worldwide sensation of terror when published in 1818. Even today, after years of being frightened by real man-made creations, including the atomic and hydrogen bombs, people can still be frightened by the thought that a laboratory-created monster might be unleashed on the world.

There may be an innate fear in us that somewhere, someday, somehow, the dividing line between human and inhuman will be broken. Alien beings will walk among us—either visitors from other planets or products of our own creation—because scientists have tampered with the stuff of life.

The world has gotten used to Frankenstein, but it is not comfortable with much of genetic engineering, and especially not with the idea of cloning humans. Psychologically, the idea of cloning human beings frightens many of us.

Scientists themselves agree that the implications of genetic engineering and cloning are so grave that they warrant immediate exploration by all levels of society. Among questions to be resolved

Left: Will cloning create monsters like the old movie Frankenstein and his bride (above) or will clones be human, like the new bride of Frankenstein?

are these: Should scientists be permitted to create human life in a laboratory? If so, who will be parents to the children that result? How will society receive these children? Will marriage as a mating institution still be necessary?

"The capacity to tinker-toy with the genetic building blocks of human beings could be used for incredible evil or incredible good," says Dr. David Roy, director of the Centre for Bioethics in Toronto, Ontario, Canada.

Some critics of genetic engineering worry that molecular biologists could become like the sorcerer's apprentice. If they allowed an altered and harmful bacterium to escape the laboratory, say these critics, it might spread throughout the environment, changing or poisoning life as we know it today.

Scientists counterargue that in calling them to account for their work with genetically modified bacteria, the public and other critics are seizing an opportunity to establish a moral principle. To scientists, the ultimate question is not whether bacteria can be contained in special laboratories but whether today's scientists can be contained in an ordinary society.

A congressional representative from California, Henry Waxman, has stated, "The question is not can new life survive outside the laboratory, but can our traditional values survive within it?"

Several years ago, the U.S. Court of Customs and Patent Appeals ruled that any corporation that creates a new form of life in the laboratory can patent it. The case on which the decision was based

Dr. Ananda Chakrabarty, who developed the oil-eating bacterium for cleaning up oil spills, believes that a toxic-chemical bug can be developed to dispose of chemical wastes.

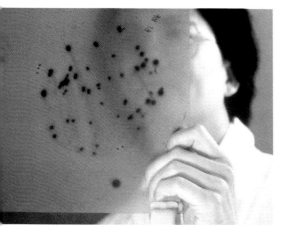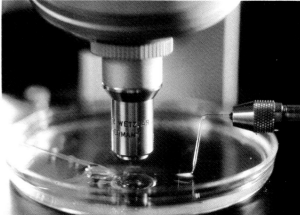

Restrictions on genetic research could mean losing very valuable knowledge. Here a scientist studies an autoradiograph (left). The dark spots on the film represent cells that contain the human gene for interferon. At the right, the genetic makeup of cells in a petri dish are studied.

involved an oil-eating bacterium for cleaning up oil spills. But many wonder if the rule also covers cloning.

Can the first person to clone a human then patent his formula? And if so, is a clone then a product, not a person? What are the clone's legal rights? Is it to be looked upon as a robot, to do the will of the master from which it was cloned? Can it be used for spare parts as has been suggested?

Even most critics agree that to restrict genetic engineering and even cloning-related research would mean closing the door on a very important area of knowledge.

DIALOGUE ON GENETICS

When a symposium on genetic engineering was held at the National Academy of Sciences in 1977, some critics held up a banner quoting Adolf Hitler saying it would be a good thing to genetically perfect the human race. Others held up signs with skull and crossbones warning of "biohazards" from genetic tampering.

In 1985, when the most recent symposium on genetics was held at the Academy, there were no such banners or signs. Mobs had crowded the meeting eight years before. This time, half the seats in the auditorium were empty. Scientists, lawyers, and industrial leaders were left in peace to review the progress of research on how to alter organisms genetically.

An overriding awareness of the public's interest in and concern for advances in genetic research was apparent, however. As one scientist put it, "If we don't treat the public as our necessary friend, none of the benefits of our research will be ours."

Those attending the symposium acknowledged that genetic experimentation raises serious questions of environmental safety and also of ethical and moral limits that genetic engineering should not exceed. Some scientists warned that their inability to predict the physical effects of inserting genes into animal germ lines dictates that such experiments should proceed with great caution.

The genetics experts agreed that what is needed is for them to meet with their critics and other concerned citizens and carry on a dialogue about the risks and ethical challenges of the new biotechnology.

GUIDELINES REGULATING GENETIC ENGINEERING

Late in 1985, the first national guidelines for human gene therapy were established by the National Institutes of Health. The institutes, headquartered in Bethesda, Maryland, form the federal government's main agency for support and conduct of medical research. The guidelines were expected to have immediate impact on research groups preparing for tests of the revolutionary new treatment and on all future genetic research in the United States.

Human gene therapy is intended to cure a patient of an inherited disease without producing effects that would be passed on to future generations. Gene therapy involves transplanting genes into a patient's cells to correct an otherwise incurable disease caused by an inborn failure of one or another gene. The treatment is considered feasible for a few rare but deadly diseases, including certain defects of the immune defense system.

In final form, the guidelines are expected to become the federal government's most concrete and explicit policy statement on gene therapy. They set the basic criteria for what ethical and scientific safeguards must be met before such new experiments are tried.

The new guidelines were prepared by a committee of scientists, lawyers, ethicists, and specialists in public policy issues. One provision of the guidelines is that government review of each application of genetic engineering be made in public before any human gene manipulation is undertaken, at least for the first round of new proposals. "We think that is essential," said the chairman of the committee that drafted the document, Dr. LeRoy Walters of Georgetown University's Kennedy Institute of Ethics.

The guidelines are considered to be a uniform set of national standards. They are defined as "points to consider" to distinguish them from the health institutes' guidelines for gene-splicing research in general. Those have been in effect since 1976.

Scientists and their critics and regulators agree that national standards are not enough. Genetic research is a worldwide pursuit and, therefore, uniform guidelines governing future research throughout the world also are needed.

A major role in understanding and appreciating the significance of genetic engineering is and will continue to be played by both print and television journalists, researchers agree. Robert C. Cowen, natural science editor of the *Christian Science Monitor,* has aptly expressed the role of the press in gene therapy: "The story of gene therapy should be reported factually and insightfully as it unfolds in the years ahead. But the dignity and the privacy of the people undergoing treatment need to be respected and protected. There is no justification for the 'media circus' atmosphere that characterizes much of the coverage of the artificial heart." Or, others add, for the sensational reportage of other organ transplants, such as in the baby girl in California whose underdeveloped heart was replaced with that of a baboon in 1984.

Genetic engineering should not be banned, most agree. It should be monitored and regulated, and the public has a right to know what is going on in genetic research and experimentation.

This vegetable is the result of genetic engineering. It is a hybrid that is the result of crossbreeding cauliflower and broccoli.

TODAY AND TOMORROW IN GENETICS
8

All systems are *Go!* in genetic engineering. Future possibilities for genetic miracles are almost limitless. If you think genetic engineering is changing the world as we know it today, wait until tomorrow—or the day after.

Early in 1985, the U.S. Department of Agriculture announced plans for increased genetic engineering and other biotechnology research at the nation's largest agricultural research center, located in Beltsville, Maryland. "The research could lead to increased animal and plant resistance to diseases, safer disposal of pesticides and other wastes, and new efficiencies for crop and animal production at lower cost," said Waldemar Klassen, director of the center.

Research is already under way or will soon be started by the center's four hundred scientists in three major categories:

△ identifying and isolating genes that could be manipulated, for example, to enable grain crops to use atmospheric nitrogen, which now is unavailable to them

△ determining the structure of cell membranes in animals and plants and then regulating the membranes for various purposes. These include preventing cold injury of plants, developing controls against parasites in animals, and controlling ripening in fruits and vegetables.

△ identifying the role of cells and antibodies so they can be manipulated to help animals resist diseases and finding out how cells communicate so that new techniques can be developed to improve animal health, reduce crop losses, shorten harvest time, and introduce insect controls that are safe for both the public and the environment

These dishes contain cells that are used in interferon studies. Those with sufficient interferon to survive a viral attack stain dark blue.

GENETICS IN HUMAN MEDICINE

Through genetic research—analyzing the tiny, programmed building blocks of life within each of us—geneticists are learning how to personalize medical statistics and medical advice to help humans live healthier and longer lives. Geneticists already can tell who has a good chance of developing sickle-cell anemia or certain types of mental retardation. Genetic research over the past ten years and ongoing studies indicate a great potential for predicting and preventing birth defects and diseases.

Most diseases have some genetic component. Some people are born with high heart-attack risks because heart trouble "runs in the family." Some people catch more colds than others. When environmental insult, such as stress or virus, meets genetic predisposition, illness results. This can be applied to cancers, infections, blood disorders, obesity, mental disabilities, and almost any other malady.

Perhaps in the next ten years, any one of us will be able to get a computer printout based upon genetic analysis, perhaps of our blood. Such a printout could tell doctors how susceptible we may be to scores of diseases, some of which susceptibility may be inherited from our parents. With such information, all of us will be able to practice our own preventive medicine.

A person with a known allergy can avoid a fruit or vegetable or weed patch that makes him or her sneeze. Similarly, a person with a known genetic defect at birth can live a more normal life by adjusting for the defect. People who are predisposed to fatty deposits in the arteries can avoid high stress by adapting their life-styles. Those missing the protein alpha-1-antitrypsin will know it is smarter not to smoke, unless they want to contract emphysema. Genetic researchers already have warned young women not to take jobs working near certain chemicals that might harm the babies they may have one day.

These commercial fermentation units are growing cultures of genetically engineered bacteria and viruses for the production of interferon and other such biological products.

RESEARCH ON VIRUSES

A virus is an infectious agent that reproduces in living cells. Geneticists today are working with some viruses that have been genetically engineered to protect people against smallpox and other diseases and also against totally different viruses that were unknown until recent years. These rebuilt viruses represent the beginning of a new vaccine strategy against disease that could have a profound effect on human health for years to come.

Vaccines are among the most potent weapons ever devised to fight infectious diseases. Experts predict there will be greatly improved vaccines against flu, cholera, and several other worldwide diseases, such as malaria and hepatitis B.

The predictions are largely based on new genetic engineering strategies, particularly gene-splicing. Knowing how to read the messages encoded in genes, scientists are able to take genes apart, add new parts to the message, and fabricate totally artificial genes that are capable of fighting diseases

FUTURE MEDICAL MIRACLES

Doctors and geneticists reported progress in using dead bones to grow live ones in 1981. Using crushed bone taken from cadavers, a way was found to induce the bodies of living persons to form new bones. The discovery could be useful in correcting birth defects, treating accident victims, and fighting dental disease.

One of the first patients in this medical research was a boy who received a nose after being born without one. The treatment was developed by a team of Harvard Medical School doctors at Children's Hospital Medical Center in Boston. So far, the team has used this material to treat forty-four patients, most of them children with birth defects that caused misshapen faces and skulls.

To make the material, human bones are crushed, minerals are removed, and the remainder is purified. The powder is then mixed with water and a paste formed that is molded into the shape of the bones doctors want to build.

A scientist holds a sample of bone material used for reconstruction and transplantation. The material, made from water and demineralized human bone, is molded into the desired shape and put into place. Natural bone then forms around it.

The crushed material does not actually become new bone by itself. When it is implanted, each speck of bone dust is surrounded by fibroblast (connective tissue) cells. Through a complicated process, the fibroblast cells change to produce cartilage, which in turn eventually becomes bone.

Meanwhile, researchers at the University of Rochester's Center for Brain Research reported the discovery that brain cells can be transplanted among animals. Researchers said that fetal brain cells from one strain of rat were transplanted into the brains of adult rats of another strain. The transplanted cells grew and functioned normally, correcting brain defects.

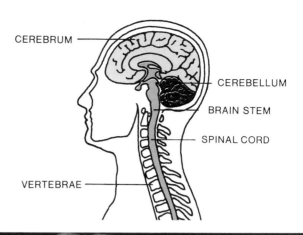

Geneticists are doing brain research that they hope will enable them to correct neurological disorders. A section of the human brain is shown here. The leaflike structure at the right is the cerebellum. Also shown are the brain stem and the cortex.

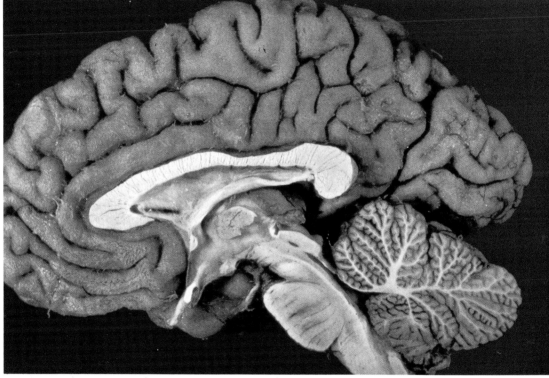

This research indicates that one day it may be possible to rejuvenate aging human brains. Damage of strokes could be reversed, memory and intelligence could be improved, and other brain defects could be corrected. Although brain-cell transplants may be years off, researchers are confident that the revolutionary discoveries eventually will have a major impact on neurological disorders.

"We are very optimistic that eventually we will be able to replace neurons [brain cells] that have become inoperative as a result of death or disease," said Dr. Don Gash.

Other geneticists are studying the possibilities of creating a "magic pill" that could prolong human life. Basic research, genetic engineering, and drugs could extend the average age well beyond ninety. The average American has already added twenty-four years to life expectancy since the turn of the century, mostly through control of infections and better nutrition. Experts on aging now believe an additional five to twenty-five years could be added through genetic advances, proper diet and exercise, giving up smoking, and avoiding drugs.

These twin doctors were ninety-two when this picture was taken. As a result of genetic research, all people may someday live to be this age or older.

One of the genetic possibilities for longer, healthier life may be an anti-aging pill. Mice given the drug appeared to be younger than those of the same age who were not given the chemical. At the same time, the drug reduced the receiver's risk of cancer. Tests are being made to find out by how much the drug will extend life. If the drug works and is safe, the human life span could be extended to 105 years.

The drug is called dehydroepiandrosterone (DHEA), a major steroid secreted by the adrenal glands. It also prevents obesity without interfering with appetite. No adverse effects from the drug have been found yet, and it may undergo human trials soon. Its first application on humans would be in women with a high risk of developing breast cancer.

MORE NEW GENETIC MARVELS

A new research library opened in 1985 offering human genes and segments of genes to scientists around the world. The gene library, operated by the Los Alamos and Lawrence Livermore national laboratories, offers a unique opportunity for investigators studying hereditary diseases.

Chemical and laser technology have simplified the process of isolating and cloning the genetic material, which may be placed in a small vial, packed into a padded envelope, and mailed to library users. Within weeks of its opening, the gene library had about two hundred requests for genetic material. The requests came from hospitals, universities, and industry laboratories.

Animal lovers who oppose the use of live animals for laboratory experiments are hopeful that new experiments will reduce the need for such guinea pigs. The Johns Hopkins Center for Alternatives to Animal Testing reported recently that researchers have been successful in growing cells, particularly skin cells, in test tubes. Artificial skin also is being developed that may replace animals in tests of drugs and cosmetic absorption.

Authorities say that many of the research projects will lead to methods that will reduce the number of animals used in laboratory testing. "The tests will not only address concerns about humane

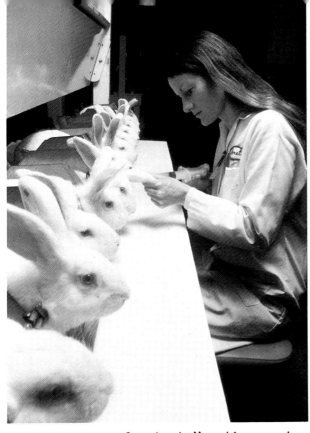

Many animal lovers object to the use of live animals such as these rabbits in laboratory experiments. New genetic techniques will enable researchers to reduce the number of animals used.

treatment of animals," said a spokesman for the hospital in Baltimore, Maryland. "They will provide laboratories with methods that are cheaper and give more information about how toxic chemicals affect humans."

GENETIC ATTACKS ON INSECTS AND BUGS

Monsanto, a chemical company, recently asked the government's approval to test the world's first genetically engineered pesticide. If it works, it could make chemical pesticides obsolete in as short a time as twenty-five years.

Gene-splicing techniques have been used to transfer the key gene from the active agent of an existing biological pesticide to a strain of common bacteria on the roots of corn plants. That gene, normally in *Bacillus thuringiensis,* directs the root bacteria to make a protein that kills insects and other pests when they try to feed on the root but that is harmless to mammals. Initial research aims at the black cutworm, which can cut corn production by 50 percent.

"Instead of having pesticide all over the field, you'd have an organism that occupies the very niche occupied by the pest," said Robert

Kaufman, Monsanto's director of plant-sciences research. "The protein attacks anything that eats the plant root."

The Japanese, meanwhile, have been conducting genetic experiments on a fungus, *Entomophaga aulicae,* which is a potent, natural enemy of the forest-defoliating gypsy moth. The U.S. Department of Agriculture's Insect Pathology Research Unit in Ithaca, New York, is hopeful that the fungus can be released in this country to become part of the natural environment and help kill the gypsy moth that is threatening our trees and forests.

Right: This bench scale fermentation unit is used to determine the exact conditions and nutrients that engineered bacteria need to grow. *Below:* Plant chloroplast cells are surrounded by *Agrobacterium tumafaciens,* plant bacteria that inject their own DNA into the cell's DNA. These are used in plant engineering.

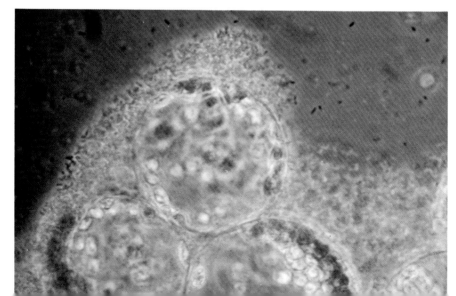

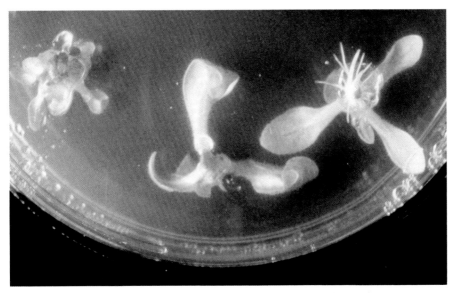

Top: Genetically engineered plantlets grown from single cells. *Bottom:* One of the 123 computer-controlled growth chambers at Monsanto's Life Science Research Center. The growth chambers, which can mimic any environmental condition, are used for biological and chemical research.

The Federal Environmental Protection Agency was considering the fungus early in 1985. If the agency gives its approval, the fungus could be tested on about half an acre of woodlands heavily infected by gypsy-moth caterpillars. Laboratory tests have been very encouraging, and the fungus appears to pose little threat to anything but gypsy-moth and several other caterpillar pests.

Another pest, the cockroach, may have a new and lethal enemy. Roaches are not only a nuisance and a health hazard; they are blamed for diminishing food stocks that contribute to famine in some Third World countries.

Chemists at Yale University in New Haven, Connecticut, have produced a synthetic cockroach aphrodisiac that kills roaches. The substance, called periplanaone-B, lures the insects into traps treated with insecticide. The aphrodisiac sends male roaches into sexual frenzies leading to their deaths. They stand on their back legs and start flapping their wings madly. After about half an hour of this hypersexual frustration, they die. Sexual attractants in nontoxic biological lures also are starting to be used in controlling gypsy moths.

SUPERBUGS

On the negative side, many physicians and scientists warn that wholesale feeding of antibiotics to livestock is breeding pools of superbugs that could produce a major health hazard to humans, as well as animals. The nation's largest epidemic of salmonella poisoning, which affected more than eighty-five hundred people in the Midwest in the spring of 1985 and may have caused several deaths, focused attention on the superbug debate.

The source of the food-poisoning epidemic, not pinpointed at the time of this writing, was believed to have been contaminated milk at a Chicago-area dairy. Since the epidemic involved an antibiotic-resistant strain of superbug, health experts warned that superbugs could cause even greater health hazards in the future.

Despite evidence tracing superbugs to livestock feeding practices, some leading scientists and organizations, including the Council for Agricultural Science and Technology, oppose a ban on feeding penicillin and tetracycline to livestock.

"Billions of food animals have received antibiotics," said Virgil W. Hays of the University of Kentucky, "but the evidence indicates little harm to humans as a result." If a ban on feeding antibiotics to

Above: The technician is using a bench scale separations process to purify proteins produced by bacteria in a bovine growth hormone. *Left:* The animals used in bovine research are carefully tended and monitored.

livestock were enacted, it is estimated that it would cost the public $3.5 billion a year or more in increased food costs.

What scientists do to our food supply can have positive or negative results. Controversy over salmonella superbugs and the practice of giving disease-resistant antibiotics to food animals is going to continue to cloud the future for genetic engineering in plants and animals until conclusive, positive solutions are found.

This plant-tissue culture technique was developed by Monsanto. Disks punched from plant leaves are mixed with engineered plant bacteria and grown on hormone-containing media. The injured cells are infected by the DNA. Callus soft tissue that forms over the cut surface will form around the edges and will eventually regenerate into whole plants.

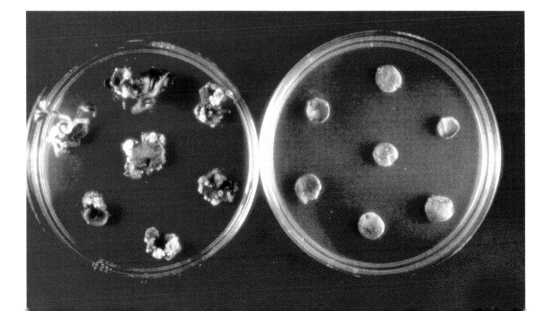

THE GENETIC FUTURE IS NOT ALL ROSY

While geneticists place a lot of emphasis on the benefits their research can offer mankind, environmentalists and others are concerned about possible adverse effects of genetically altering life in plants, animals, and humans.

One worry is that however harmless a new genetically engineered variety may appear to be, it could wreak ecological havoc simply by outrunning other species. Opponents of unregulated experiments compare the potential hazards to those caused by the water hyacinth, a weed that was unintentionally introduced to this country from South America some years ago. Now the plant chokes many important waterways, endangering plant and animal life.

Environmental Policy Institute spokesmen warn that, for example, herbicide-resistant plants might promote use of toxic substances, which have already contaminated groundwater and threatened public health in some agricultural areas.

A 1984 court case began studying adverse effects of an experiment that would have used genetically engineered bacteria to keep potatoes free of frost. The Foundation for Economic Trends argued that using the hardier crop would upset the ecological balance. Ultimately, the bacteria might alter the climate.

The colonies of bacteria being grown on these nutrient plants will be used with DNA technology to help find ways to design superior plants to combat the growing world food shortage.

Some blame a worldwide increase in malaria on the Green Revolution of genetic engineering. Anthropologists in 1983 argued that genetically engineered crops introduced in some Third World countries produced large harvests but that these hybrid plants were weak and had to be protected by the extensive use of pesticides. The crop spraying reportedly led to a more dramatic rise in the resistance of a strain of mosquitoes to toxins than any previous use of insecticides to control the disease. A serious outbreak of malaria followed.

"The origins of this major ecological disaster must be sought as much in the unwitting actions of international groups such as the World Health Organization and the Food and Agricultural Organization as in hapless nature," said Robert Wasserstrom, a Columbia University anthropologist.

To avoid such ecological disaster, sound scientific pest-management strategies that couple limited pesticide use with other techniques that don't interfere with public health efforts are recommended. Agricultural officials, health authorities, and geneticists need to work together to control crop pests and disease.

Similarly, geneticists and pharmacologists are discovering that the common practice of prescribing standard doses of drugs to all patients may be wrong and potentially harmful. Growing evidence indicates that each person's body reacts differently to drugs. For many patients, a standard dose could be dangerous; for others, it may be ineffective. These differences are genetically determined so that what is safe for one person may be dangerous for another.

A new field called pharmacogenetics is attempting to understand the genetic variations that govern individual drug responses. With enough knowledge, adverse drug reactions could be predicted and avoided. Such research also is designed to improve drug therapy.

WHERE DOES GENETICS GO FROM HERE?

As has been said, the future for genetic research is almost limitless. Almost every day, in newspapers and magazines and on television news programs, we learn of new and often astounding discoveries in genetic engineering of plant, animal, and human life. The aim is to improve species so they can be healthier and more productive. Whether this will include human cloning is still an arguable question.

At the time this book was being written in the spring of 1985, researchers predicted that in a year or two, geneticists would be able to insert a new gene into a child's cells in an effort to cure a devastating disease. Medicine would thereby enter a new and controversial era—the age of gene therapy.

Gene therapy is the alteration of the genetic material within human cells to cure or prevent inherited diseases. The first efforts at gene therapy involve somatic (nonreproductive) cells, so that any genetic changes will not be passed on to the children of the person genetically changed. Because of its widespread implications, the federal govern-

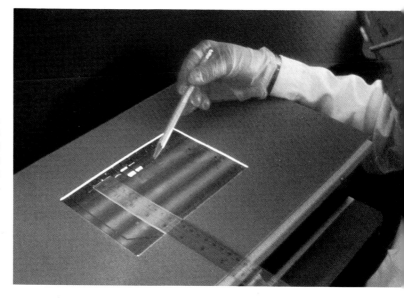

A scientist examines DNA fragments separated on an electrophoresis gel. The bands become visible under ultraviolet light when a dye is bound to DNA.

ment, the researchers involved, and some public-interest groups are giving gene therapy careful scrutiny for safety and ethics.

Three other major advances in genetic research were revealed as this book was being prepared. First, plant geneticists at the Morton Arboretum near Lisle, Illinois, reported that they had successfully developed a tree that is resistant to Dutch elm disease. The new tree looks much like the American elm but will be able to resist the deadly disease that has claimed more than forty million American elms in the last fifty years.

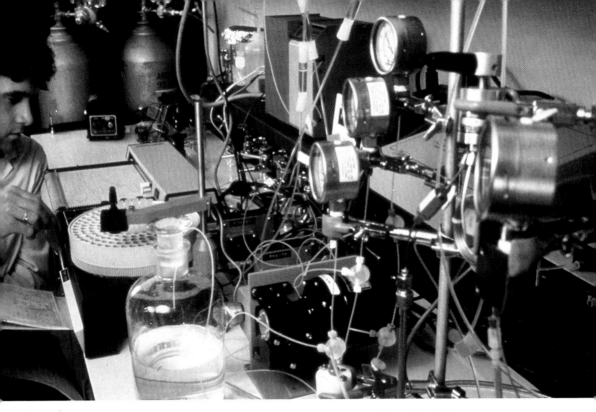

Scientists are trying to learn more about the cellular protein factories. This researcher is separating and purifying cell proteins, using a technique called high performance liquid chromatography.

Second, researchers at the Neuropsychiatric Institute of the UCLA Medical School reported that autism, a severe mental disorder that becomes evident in infancy, seems to have a genetic cause. Studies indicate that autism is associated with an inherited gene and that the pattern of inheritance is recessive. The new finding argues strongly against theories that autism is primarily a result of behavioral factors such as interactions between parents and infants. Now that scientists are certain that pathogenic genes cause autism, the next step is to determine where they reside on the gene map.

Finally, a Swedish scientist reported that he had reproduced in bacteria segments of DNA isolated from an Egyptian mummy dating back two thousand years. Analysis of one of the segments shows that it suffered little damage over the two millennia of preservation.

This finding and the recent cloning of DNA from a quagga (see page 62) raise the hope that recombinant DNA techiques may one day be applied systematically to archaeological samples. The interaction of DNA technology and archaeology may open a new phase in our understanding of human history.

But the Swedish scientist is quick to assure us that though it may be possible to clone pieces of chromosomal DNA from biological remains of a two-thousand-year-old mummy of a child, "it is a research approach, not a mummy, that is coming to life. From this preliminary research, we cannot of course reconstitute a functional gene, let alone a living individual."

It is too great a temptation not to bring this look at genetic research to a close on a note of both amazement and humor by reporting what the Chinese are doing. A few years ago, Chinese scientists experimented with fertilizing a chimpanzee with human sperm. The idea was to attempt to create a "near-human ape." The creatures, said the Chinese, could be used for herding sheep and cows and for driving carts. They also could be used in exploring space, exploring the bottom of the sea, and in mines.

The primary object of the experiments was to develop a creature with a larger brain and mouth. One factor that makes it impossible for a chimp to imitate human sounds is that its mouth is too narrow.

The research was spurred by the evolutionary belief that humans and apes were originally related and still are compatible enough to produce common offspring. It is a point that has led to much fantasy and humor in the Western world.

The Chinese succeeded in fertilizing a chimp with human sperm, but when the female was three months pregnant, the experiment was halted. Reports out of China indicate that the experiment may someday be tried again.

Though there are shades of Frankenstein's monster in the idea of birthing a superape, Chinese geneticists maintain that the creation could benefit humanity. "My idea is that man is the master of the natural world," said Dr. Ji Yongxiang of the Chinese Academy of Science in Peking. "Although the human population is growing all the time, this doesn't preclude the possibility of developing other species. A near-human ape could serve the needs of mankind."

Supercows? Superapes? Cloned humans? It could all happen tomorrow in the amazing world of genetics.

GLOSSARY

bacterium. A single-celled microorganism with a primitive nucleus.
chromosome (colored body). A piece of DNA containing many genes. Chromosomes are the location of hereditary (genetic) material within the cell.
clone. One or more genetically identical organisms.
DNA (deoxyribonucleic acid). The molecule containing hereditary information in all but the most primitive organisms. The stuff of which genes are made, a long chainlike molecule composed of nucleotides.
gene. The part of a DNA molecule that comprises the basic fundamental hereditary unit. Genes contain the coded information for making a specific protein.
gene modification. A process of genetic therapy in which genes are altered in the living organism. It is not yet possible but is expected in the future.
gene transplantation. A technique of moving an entire gene from one organism into another.
genetic engineering. The science of working with genes to study and possibly alter and thereby improve life forms.
genetics. The science of heredity dealing with resemblances and differences of related organisms resulting from the interaction of their genes and the environment.
heredity. The transmission of genetic characteristics from parents to offspring.
implantation. The process by which the fertilized egg (zygote) becomes attached to the wall of the uterus, which then serves to nourish the embryo through growth and subsequent development.
in vitro (in glass). Outside the body, in the laboratory, in the test-tube.
protein. A molecule of linked amino acids. Proteins serve primarily as biochemical catalysts and as structural parts for an organism.
recombinant DNA. A new combination of genes spliced together on a single piece of DNA.
zygote. A fertilized egg. A product of the fusion of sperm and egg.

MORE ABOUT GENETICS

Asimov, Isaac. *How Did We Find Out About Genes?* New York: Walker & Co., 1983.
Bornstein, Jerry and Sandy. *What Is Genetics?* New York: Julian Messner, 1979.
Cherfas, Jeremy. *Man-Made Life—An Overview of the Science, Technology, and Commerce of Genetic Engineering.* New York: Pantheon, 1983.
Facklam, Margery and Howard. *From Cell to Clone—The Story of Genetic Engineering.* New York: Harcourt, Brace, Jovanovich, 1979.
Morrison, Vera Ford. *There's Only One You—The Story of Heredity.* New York: Julian Messner, 1978.
Silverstein, Alvin and Virginia. *The Code of Life.* New York: Atheneum, 1972.
———. *Futurelife, the Biotechnology Revolution.* Englewood Cliffs, N.J.: Prentice-Hall, 1982.
Sylvester, Edward J. *The Gene Age—Genetic Engineering and the Next Industrial Revolution.* New York: Scribner's, 1983.

INDEX

agrigenetics, 29, 34
AIDS, 72-73
allergy, 111
amniocentesis, 80, 87
animal genetics, 41-63
antibiotics, 120-121
aphrodisiac pesticide, 120
archaeological samples, 18, 60-62, 125-126
artificial insemination, 17, 46, 57, 83-89, 90
artificial implants, 19, 116

back-breeding, 61
bacteria, 11, 30, 104, 111, 117, 118, 122
bioengineering, 21, 34, 105
bioscientists, 13, 27, 109
birth defects, 80, 110-111, 125
blindness research, 76-77
blood cells, 8, 10, 71
bone building, 112, 113
brain, 19, 20, 21, 64, 113-114

cancer, 10, 12, 65, 73, 78-79
cell transplants, 19-21, 78, 113-114
cells, 6, 7-8, 30, 31-32, 49-51, 63, 6, 69, 113-114, 118, 124, 125
chemical wastes, 45, 104
chimeras, 50-51
chromosomes, 6, 8, 54, 60, 63, 74, 99-100
cloning, 17-18, 22, 73, 78, 97-100
cloning animals, 41, 48-55, 62-63
cloning drugs, 67
cloning humans, 21-23, 95-102, 103-106
cloning plants, 23, 26-27, 29, 36
computer, 7, 21, 38, 111
corn plant, 26, 27

diabetes, 8, 71
dialysis, 75, 76
diseases, 58, 70, 110-112, 120-121
   See also genetic diseases
DNA (deoxyribonucleic acid), 6-7, 11, 18, 60-62, 68, 69, 70, 81, 98, 101, 124, 125-126
DPT, 75
drugs, 67, 127
dwarfism gene, 31

eggs, mammal, 52-54, 93
embryo splitting, 46, 47, 48, 99
embryo transplant, 43-44, 45, 57, 83-90, 98-99
endangered species, 45-46
environmental disasters, 45-46
enzymes, 29, 65, 68
ethics, 13, 17-18, 23, 90-93, 97-100, 103-106, 124

evolutionary process, 19
eye diseases, 76

fallopian tubes, 14, 15, 83-84
farm animals, 42-43, 56-59
fertilized egg, 52-55, 61, 63, 83, 85, 87-88, 98-100, 126
food poisoning, 120-121
food supply, 13, 32
freezing embryos, 88-89
fungi, 54, 118-120

gene implants, 66, 77-78
gene location, 69, 74
gene sequencing, 72
gene-splicing, 27, 29-31, 58, 112, 117
gene therapy, 67-69, 78, 106-107, 124
genetic code, 68, 77
genetic diseases, 9, 12, 65-69, 73-78, 80, 87, 110, 125
genetic engineering, 9-23, 41, 56-62, 65, 68, 81, 90-93, 103-107, 122-123
genetic research, 5, 38, 80, 109-126
genomes, 63
germ cells, 38, 98-99
germination, 38-39
Green Revolution, 28, 38, 123
growth disorders, 11
growth hormone, 42, 58, 69, 81, 120, 121

herbicides, 31, 37, 122-123
hereditary diseases, 12, 69, 116
heredity, 5, 9, 67
hormones, 8, 42, 85, 87
human genetics, 65-82
hybrid plants, 15, 16, 17, 25, 27, 28, 123

immune system, 70-72, 78
implants, 19, 66, 77-78, 98-99
in vitro fertilization, 14, 48, 63, 82, 83, 84, 85, 86, 87-93
infectious diseases, 112
infertility, 83
insects, 18, 60-61, 118-121
insulin, 8, 12, 69, 71
interferon, 65, 110
interleukin, 67
isopentenoids, 31-32

kidney disease, 75-76

laboratory animals, 55, 117
laparoscope, 85
law and genetics, 89, 90, 104, 107
leukemia, 10, 79
lippia dulcis, 35

living insecticide, 31

malignant cells, 79
medicine and genetics, 110-116
meiosis, 63
mutation, 100

neurons, 114
nuclear transfer, 49, 51, 53

parthenogenesis, 53
periplanaone-B, 120
pesticides, 45, 117-118, 123
phamacogenetics, 123
plant bacteria, 118, 122
plant breeding, 25-29, 33
plant cells, 32, 109
plant cloning, 23, 26-27, 29, 36
plant genetics, 34-35, 38, 124
plant hybrids, 15, 25, 27, 34, 123
prescription drugs hazards, 123
proteins, 67, 68, 70, 71, 111, 117-118, 125

quadruplets, 82

recombinant DNA, 9-11, 12, 25, 29, 58, 68, 71, 74
recombinant TNF, 79
regeneration, 19, 20-21
ribonucleic acid (RNA), 68

salmonella, 120-121
selfing, 50-51
sex prediction, 14, 57
sickle-cell anemia, 9, 12, 69, 110
single cell, 50, 66
skin cells, 116
skull, 64
sperm, 83, 91, 93, 98-99
superanimals, 13, 41-64, 126
superbugs, 120-121
superchickens, 58-59
superplants, 15-16, 25-40
surrogate mother, 43-45, 46, 52, 53, 56

T-cells, 70, 71, 73, 79
test-tube babies, 14-15, 44, 83-93
tissue culture, 32-34
transplants, organs, 20, 76
tree research, 23, 36, 36, 37
triplets, test-tube, 88
twins, 46, 47, 48-49, 63, 87, 88, 94, 95, 115

vaccines, 11-12, 58, 112
viruses, 72, 73, 78, 79, 112

zebra, 43, 45
zygote, 49

About The Author

Walter Oleksy, a Chicago area free-lance writer, is the author of forty-four books, most of them for children. A former newspaper reporter and magazine editor, Oleksy, a bachelor, lives in Evanston, Illinois, with his fourteen-year-old Labrador.